KNOWLEDGE, CAUSE, AND ABSTRACT OBJECTS

KNOWLEDGE, CAUSE, AND ABSTRACT OBJECTS

Causal Objections to Platonism

by

COLIN CHEYNE
University of Otago,
Dunedin, New Zealand

KLUWER ACADEMIC PUBLISHERS
DORDRECHT / BOSTON / LONDON

A C.I.P. Catalogue record for this book is available from the Library of Congress.

ISBN 1-4020-0051-0

Published by Kluwer Academic Publishers,
P.O. Box 17, 3300 AA Dordrecht, The Netherlands.

Sold and distributed in North, Central and South America
by Kluwer Academic Publishers,
101 Philip Drive, Norwell, MA 02061, U.S.A.

In all other countries, sold and distributed
by Kluwer Academic Publishers,
P.O. Box 322, 3300 AH Dordrecht, The Netherlands.

Printed on acid-free paper

Printed in the Netherlands.

For Elizabeth

TABLE OF CONTENTS

FOREWORD

The main purpose of this book is to argue against the claim that abstract entities exist, where abstract entities are entities that lack causal powers and are not located anywhere in space and time. My particular target is mathematical platonism. Mathematical platonism is the view that the subject matter of pure mathematics is abstract mathematical objects. Accordingly, mathematical platonism incorporates the view that our mathematical knowledge includes knowledge of the existence and properties of abstract or platonic objects. In outline, my argument against mathematical platonism is that, since it is in virtue of their causal powers that we come to know of the existence of any entity, we cannot know of the existence of causally impotent objects and, accordingly, we ought not to believe that such objects exist. It follows that we ought not to believe mathematical platonism and the same applies to any other platonist theory that appeals to the existence of abstract objects.

I have a number of aims for my project. My overall aim is to present the strongest causalist case that I can against platonism. Although anti-platonists often appeal to causal objections, few detailed accounts of such objections have been made out. So realising this aim will, at the very least, fill a gap in the literature on epistemology and the philosophy of mathematics.

A number of platonists have argued that any account of knowledge that is strong enough to rule out knowledge of abstract objects will be too strong, in that it will also rule out knowledge that we clearly have. So one of my aims is to present an account of knowledge that avoids this criticism. I offer a necessary condition for knowledge that does not allow platonic knowledge but does not disallow any of the uncontroversial examples of knowledge that defenders of platonism have raised. Now a constraint on knowledge that simply deals with these purported counterexamples may be quite implausible or *ad hoc*. However, I also hope to convince the reader that my account of knowledge is plausible and, indeed, that it is superior to other competitors.

Note that what is crucially at issue is our knowledge of the *existence* of certain entities. The fact that the relevant causal constraint need only apply to existential knowledge has often been overlooked. So I argue for a plausible constraint on our existential knowledge, one that rules out our having knowledge of the existence of platonic objects without ruling out our ability to know of the existence of other, less controversial, entities.

Even if I do not succeed in convincing the reader of the merits of my causal constraint on existential knowledge, there is a further plank in my case against platonism. Platonists owe us an account of how it is that we have mathematical knowledge, supposing that knowledge to be platonic knowledge. In the latter part of the book (Chapters 8-12) I examine several accounts of platonic knowledge and conclude that none is satisfactory.

If platonism is to be discarded, then the question arises as to what should take its place. If not abstract objects, then what is the subject matter of mathematics and the other discourses that have been given platonistic analyses? Answering this question is not one of the aims of this book. Here I simply observe that most arguments for platonism are fundamentally linguistic. They take the nature and structure of our mathematical (or some other) language to be a guide to the nature and structure of mathematical (or some other) reality. This, in itself, may be no bad thing. But we need to be cautious, lest we get the linguistic cart before the metaphysical horse. After all, we can speak of the very same fact by employing different items of language with entirely different structures. Which structure is to have priority? When problems arise, such as the epistemological problems discussed in this volume, we need to look beyond our language to the world itself, especially to our mathematical (or other) practices and to the ways in which we use our language.

My project is epistemological. The enterprise that epistemologists are engaged in is (or ought to be) directed fundamentally at answering the questions, 'What can we know?' and 'What ought we to believe?' My modest project contributes to this greater enterprise by concluding that we cannot know of the existence of abstract objects and that we ought not to believe that such objects exist. In his autobiography, John Stuart Mill wrote:

> The notion that truths external to the human mind may be known by intuition
> or consciousness, independently of observation and experience, is, I am per-
> suaded, in these times, the great intellectual support of false doctrines and bad
> institutions...There never was a such an instrument devised for consecrating
> all deep-seated prejudices. (Mill 1873/1992, p. 123)

Insofar as Mill refers to truths with existential import, I support his senti-ment. Mill has in mind the pernicious effects of this 'false philosophy in morals, politics, and religion' (*ibid*). Confined to pure mathematics and the like, it may be a harmless curiosity. Nevertheless, by repudiating mathematical platonism, I believe we move the epistemological enterprise in a direction that has the beneficial effect of undermining other ill-founded and more damaging doctrines.

ACKNOWLEDGEMENTS

This book has its origins in an undergraduate essay that I wrote for a course on the philosophy of mathematics. The topic of that essay was broadened to become the topic of my PhD thesis (Cheyne 1994). In turn my thesis was developed to provide the basis of this book. The lecturer who sparked my initial interest, the supervisor who sustained and nourished that interest, and the head of department who ensured a research environment in which this book might come to fruition was, in all cases, Alan Musgrave. I am deeply indebted to Alan for his enthusiasm, encouragement, and invaluable assistance. Any reader should also be grateful to Alan. By rescuing me from countless stylistic infelicities, he has made reading the book a less unpleasant task than it would otherwise be. Any mistakes that remain are, of course, my own bloody fault.

It has been my good fortune to profit from the example and support of my colleagues, past and present, in the Department of Philosophy at the University of Otago. I am particularly grateful for numerous discussions with Charles Pigden and the inspiring example of the late Pavel Tichý. I also acknowledge and express my thanks for the help and suggestions of Mark Balaguer, JC Beall, John Bigelow, John Bishop, Phillip Catton, Mark Colyvan, Peter Entwisle, Hartry Field, Alvin Goldman, Paul Griffiths, Bernard Linsky, Robert Thomas, and Ed Zalta. I should like to mention two books that were particularly influential, especially in the early stages: Andrew Irvine's edited volume for this series *Physicalism in Mathematics* and Hartry Field's *Realism, Mathematics and Modality*.

I am grateful for the encouragement and assistance of the series editor, Bill Demopoulos, and of Kluwer's editorial staff, Rudolf Rijgersberg and Jolanda Voogd. A special thank you to David Talbot for his advice and help with research and computing, and to Penny Bond and Sally Holloway for their invaluable clerical services.

My heartfelt gratitude to Elizabeth for her support and encouragement, when even grudging tolerance is more than one should expect for such an unwarranted intrusion into a marriage relationship, and to Casimir, Bijou, and Tybalt for their important contribution to the maintenance of my sanity.

The preparation of the manuscript was partly funded by a research grant from the Division of Humanities, University of Otago, for which I gratefully acknowledge receipt.

In places I draw on and incorporate material from previously published articles. I thank the following publishers for their permission to reproduce material from those articles: Núcleo de Epistemologica e Logica for Cheyne 1997a, the International Phenomenological Society for Cheyne 1997b, the Australasian Association of Philosophy and Oxford University Press for Cheyne 1998 and Cheyne & Pigden 1996, the International Phenomenological Society for Cheyne 1997b, and R.S.D. Thomas and the Canadian Society for History and Philosophy of Mathematics for Cheyne 1999.

Felix qui potuit rerum cognoscere causas.
Virgil

Ah, la belle chose que savoir quelque chose.
Molière

There is occasions and causes why and wherefore in all things.
Shakespeare

CHAPTER 1

PLATONISM AND CAUSALITY

1.1. PLATONIC OBJECTS

For more than two thousand years, philosophers have postulated the existence of objects that exist outside space and time. Plato argued that as well as the particular objects given to us in sense experience, such as human beings, cats, beds, and cartwheels, there exists a quite different type of entity, which he called Forms or Ideas. Examples of Plato's Forms are Beauty, Circularity, Whiteness, Humanity, and Cathood. These Forms are what are supposedly designated by predicates in subject-predicate sentences. The Form of Whiteness is designated by the predicate in 'My cat is white'. Forms are what account for the sameness-of-property of particulars. It is in virtue of their participation in the Form of Humanity that Socrates, Confucius, Geronimo, and Sylvester Stallone are all human. According to Plato, the Forms are eternal, unchanging, and have no location in space or time.[1]

There is a long tradition of regarding numbers and other mathematical entities as existing unchanging and without spatio-temporal locations. Once again, the tradition can be traced back to Plato.[2] A typical argument for this view goes as follows. Statements such as:

---

[1] Plato's theory of Forms is scattered throughout most of his writings; see in particular *Parmenides* 129-35, *Phaedo* 65-66, 74-79, 100-02, and *Republic* 476-80, 507-18, 595-96. Specific reference to the Forms' unchanging nature is in *Phaedo* 78-79, to their non-temporality in *Timaeus* 37-38, and to their non-spatiality in *Phaedrus* 247. Plato's theory is summarised in Staniland (1972 pp. 1-7) and in Wedberg (1955 ch. 3).
[2] See in particular *Phaedo* 103-05, *Republic* 510, 525-27, *Parmenides* 129-30, *Theaetetus* 185.

1

(A)   Helen Clark is a prime minister

are true in virtue of the existence of a certain object possessing a certain property. Likewise, statements such as:

(B)   Five is a prime number

must be true in virtue of the existence of a certain object (viz. the number five) possessing a certain property. Since (B) is indeed true, the number five does exist. But we do not suppose that the number five is located in any particular place. Any number of events could bring it about that (A) is no longer true, but no matter what the world is or could be like, we expect that (B) would still be true. Thus, the number five must always exist and do so without changing in any way. It also follows that its existence is mind-independent. (B) is true no matter what we think or believe or desire.

Other candidates for being mind-independent, non-spatio-temporal objects are propositions, meanings, truth-values, universals (properties and relations), locations, possible worlds, and so on.

There are reasons for supposing that if such objects exist then they lack causal powers and causal properties, or at least the causal power to influence human beings.[3] We usually suppose that in order to exert a causal influence, an object must do so at some particular time and place, and this would not be possible for an object lacking a spatio-temporal location. It is usually supposed that the cause of an event is another event and that all events involve change of some sort. Objects which are unchanging and unable to be influenced in any way by the activities of the spatio-temporal world cannot enter into causal relations with spatio-temporally located objects and, thus, could neither causally influence nor be influenced by them. If we believe that human beings are completely located within space and time, then it seems that we must be totally isolated from objects that exist outside space and time. Even if such objects possess causal powers, it does not seem that their powers could influence us in any way.

It is widely held, then, that objects exist that are outside of space and time, and are causally isolated from human beings. I shall refer to such objects as 'platonic' objects. I acknowledge that some philosophers endorse the notion that so-called platonic or abstract objects do have causal powers

---

[3] Some may object to the notion of causal powers, for example, adherents to a Humean regularity thesis of causation. Such persons can regard talk of entities with causal powers as a *façon de parler* for entities that can enter into causal relations or entities for which causal claims may be true.

and properties and do have spatio-temporal locations.[4] However, my use of 'platonic' is not intended to encompass such objects. Only the theories of those who agree that their posited entities are acausal, come within the purview of my discussion.

Ontological platonism is the claim that platonic objects exist. Epistemological platonism is the claim that human beings can and do have knowledge of the existence and properties of such objects. Since knowledge entails truth (see Appendix I, Section 2), epistemological platonism entails ontological platonism. To affirm ontological platonism while denying epistemological platonism would be untenable, even if it is not strictly contradictory. If I assert details of the inhabitants of a distant planet but deny that I have any knowledge of those aliens, then there is no reason why my assertions should be regarded as anything more than idle fancies. By 'platonism' I shall mean the conjunction of ontological and epistemological platonism. For the most part I shall address mathematical platonism, the claim that mathematical objects are platonic objects of whose existence and properties human beings can and do have knowledge.

1.2. THE CAUSAL OBJECTION

Platonists claim that we can know of the existence of platonic objects and discover various of their properties. For example, Michael Resnik holds that 'mathematical objects are causally inert and exist independently of us and our mental lives [however] we can refer to such alien creatures and acquire knowledge and beliefs about them' (1990, p. 41).

There is a long-standing epistemological objection to such claims. The argument is as follows. We can only know about something if we causally interact with it. But we cannot causally interact with platonic objects. Therefore, even if they exist, we cannot know about platonic objects.

From that conclusion, it may be further argued that we ought not to believe in the existence of platonic objects. Or (a stronger conclusion) that we ought to disbelieve in the existence of such objects. Hartry Field argues for this stronger conclusion by means of an analogy:

> We also can't have direct evidence against the hypothesis that there are little green people living inside electrons and that are in principle undiscoverable by human beings; but it seems to me undue epistemological caution to maintain agnosticism rather than flat out disbelief about such an idle hypothesis. (1989, p. 45)

---

[4] For example, Maddy (1980) & (1990), and Bigelow (1988).

But whether we ought not to believe in platonic objects or to disbelieve in them, in either case, platonism is untenable. In fact, the conclusion that we cannot have knowledge of platonic objects is sufficient for epistemological platonism to be rejected, irrespective of normative conclusions as to what we ought or ought not to believe.

Arguments against platonism based on causal grounds can be traced back almost as far as arguments for platonism. Sextus Empiricus argued against the Stoic notion of a proposition on such grounds.[5]

> If, then, [propositions] exist, the Stoics will declare that they are either corporeal or incorporeal. Now they will not say that they are corporeal; and if they are incorporeal, either—according to them—they effect something, or they effect nothing. Now they will not claim that they effect anything; for according to them, the incorporeal is not of a nature either to effect anything or to be affected. And since they effect nothing, they will not even indicate and make evident the thing of which they are signs; for to indicate anything and make it evident is to effect something. But it is absurd that the sign should neither indicate nor make evident anything; therefore the sign is not an intelligible thing, nor yet a proposition. (Sextus 1936, pp. 374-77)

1.3. BENACERRAF'S ARGUMENT

In contemporary philosophy, the *locus classicus* of causal objections to mathematical platonism is Paul Benacerraf's 1973 paper. He claims that:

> If, for example, numbers are the kinds of entities they are normally taken to be, then the connection between the truth conditions for the statements of number theory and any relevant events connected with the people who are supposed to have mathematical knowledge cannot be made out. (1973, p. 414)

He argues for this claim on the grounds that he favours 'a causal account of knowledge on which for X to know that S is true requires some causal relation to obtain between X and the referents of the names, predicates, and quantifiers of S'. He notes, in particular, Alvin Goldman's 1967 paper, which contains the first detailed causal theory of knowledge.

Because Benacerraf regards the platonist account as providing the best account of the truth conditions for mathematics, he sees his epistemological position as creating a puzzle. Many other philosophers of mathematics have accepted an epistemology similar to Benacerraf's and taken this to be

---

[5] Since much of Sextus's work is known to be unoriginal, it is probable that the argument considerably predates the second century A.D.

grounds (although not always the only grounds) for rejecting platonism and seeking an alternative account of mathematics.

Geoffrey Hellman (1989) identifies one of the desiderata of a philosophical interpretation of mathematics as requiring 'a reasonable integration of mathematical knowledge with the rest of human knowledge, something that the traditional platonist interpretations have difficulty providing'. Although he concedes that the puzzle as to 'how it is that the posited abstract objects "play any role"—"make any difference"—in our knowledge and in our language' may be a 'pseudo-problem', he concludes that '[s]urely, it would be desirable to understand mathematical discourse in such a way that such questions are completely blocked from the start' (pp. 3-4).

Philip Kitcher (1984) endorses Benacerraf's epistemological view. He claims that 'if we adopt an enlightened theory of knowledge, we should hold that when a person knows something about an object there must be some causal connection between the object and the person' and concludes from this that platonism is mistaken. Although he recognises that platonists have responded with arguments which attempt to show that our 'best theory of knowledge' does not preclude knowledge of platonic objects, he contents himself with the 'brief, dogmatic evaluation [that] the intuition behind [Benacerraf's] point is deep enough to enable it to be reformulated so as to cause difficulty for the platonist responses which have been offered to the original version' (pp. 102-03).

According to Charles Chihara (1990), by adopting traditional platonism:

> [w]e seem to be committing ourselves to an impossible situation in which a person has knowledge of the properties of some objects even though this person is completely cut off from any sort of causal interaction with these objects. ... [H]ow does the mathematician discover the various properties and relationships of these entities that the theorems seem to describe? By what powers does the mathematician arrive at mathematical knowledge? (p. 5)

Having rejected the attempts of platonists (in particular, Quine and Gödel) to provide plausible answers to these questions, Chihara suggests constructing 'another kind of mathematics that will avoid those features of the original system that gave rise to the puzzles' (p. 23).

David Papineau (1988) argues that:

> [t]he beliefs we acquire about the natural world tend to be true because in general we are caused to form such beliefs by their truth conditions. [However, t]his story doesn't work...for 'platonist' mathematical beliefs that refer to abstract objects. ...The trouble then is that we can't happily view ourselves as responsive to such truth conditions. If there are platonist mathematical facts, they exist outside the causal world of space and time. But we

are causal beings inside the spatio-temporal world. So there is no possibility
of mathematical facts exerting a causal influence on our beliefs. (p. 152)

In arguing for his physicalist theory of mathematics, John Bigelow
(1988) states that '[i]n epistemology, [his] tastes run...towards causal
theories of knowledge [which] tell us that, in order to know about something, there must be some sort of appropriate causal network linking you
with that thing.' He then claims that the causal theory of knowledge should
pose no threat to his particular brand of mathematical realism, because
according to his theory, numbers and the like are not causally impotent (pp.
175-76).

1.4. REFERENCE AND CAUSALITY

There is another objection to platonism stemming from the acausality of
platonic objects. Benacerraf alludes to but does not develop this objection
when he says that he believes (in addition to a causal account of knowledge)
'in a causal theory of reference, thus making the link to [one's] saying
knowingly that S *doubly* causal' (1973, p. 412, his emphasis). According to a
causal theory of reference, names of objects and kinds can only successfully
refer if there exists a causal-historical chain leading back from the speaker to
the object itself or a sample of the kind (Kripke 1980, Devitt 1981). But if
the objects are acausal, then no such chain is possible. Even if platonism is
true, it would seem that we cannot successfully talk about those objects
which occupy the platonic realm.

The causal theories of reference were devised as an alternative to
description theories of reference. But 'pure' causal theories are subject to
telling objections. It seems that we can fix the reference of terms by means
of a description, so the causal theory does require some modification. What
the causal theorist will insist on is that the terms used in a reference-fixing
description must ultimately have *their* reference fixed by causal chains. Thus
I can fix the reference of 'Wendy' by means of the description 'my cat's
mother' without having an appropriate causal connection with Wendy so
long as my use of the terms 'my cat' and 'mother' do have appropriate
causal connections (Devitt & Sterelny 1987).

If a descriptive-causal theory of reference is correct, then might it not be
possible to fix the reference of terms for platonic objects by means of
descriptions employing terms whose reference has been fixed by respectably
causal means? It is not clear that terms such as 'acausal' and 'non-spatio-
temporal' can be causally grounded. On the other hand, is not clear that they

cannot. In other words, it is not clear that the descriptive-causal theory of reference rules out the possibility of reference to platonic objects.

Michael Resnik (1990, pp. 47-53) suggests an alternative theory of reference. Put simply, he claims that for any **x**, a singular term 't' refers to **x** just in case **x** = t. For example, 'Duke Ellington' refers to the composer of *Black and Tan Fantasy* since Duke Ellington is the composer of *Black and Tan Fantasy*. Similarly, Resnik can claim that if platonism is true, then '2' refers to the only even prime because 2 is the only even prime.

It is not clear how Resnik's theory differs from a description theory of reference. Certainly more can be said on the relative merits of these theories. However, at this point I shall simply state that although it may be possible for us to refer to objects that are causally isolated from us but that it does not seem possible that we can have knowledge of such objects. For example, we can speculate about life on distant planets and devise precise terms for the objects of our speculations. Let us give the name 'Throg' to the leader of the green-faced, four-armed people who live on the only inhabited planet in the Ursa Minor galaxy. If, perchance, such a person exists, then we have a case of successful reference, but so long as we are unable to make contact, however indirect, with such a person, then we cannot know of its existence. It is this thought that causal isolation denies the possibility of knowledge that motivates this project.

1.5. PLATONIST RESPONSES

Platonists have responded to the epistemological objection to platonism in two ways. There is the negative response of attempting to show that the Benacerrafian argument is unsound, and the more positive response of giving an account of how it is that we come to have platonic knowledge. Both of these responses are examined in this book.

John Burgess is a platonist who has criticised Benacerraf's argument. He says of Benaccerraf's original version that 'though it may be sketchy, [it] still is the most detailed in the literature', and that '[a]n argument that is thus cited as powerful motivation by many, but defended in detail by none, invites examination' (Burgess 1990, p.3). The challenge implicit in Burgess's statement is the starting point for this book. My aim is to provide a detailed defence of the epistemological objections to platonism.[6]

---

[6] I note that Burgess has, more recently, made a close examination of such causalist arguments himself. See Burgess & Rosen (1997).

1.6. FURTHER PRELIMINARIES

As already noted, most of my discussion is in terms of mathematical objects, which are usually regarded as paradigmatic platonic objects. On a narrow reading, my arguments will constitute a rejection of mathematical platonism on epistemological grounds. I believe that the same arguments apply just as effectively to all platonic entities, so this book may also be read as an attack on all varieties of platonism that postulate the existence of acausal entities.

A distinction may be made between entities that are particulars or complete individuals having determinate identity conditions, and other entities that are not particulars or are in some way incomplete.[7] The term 'object' is sometimes reserved for the former. It may be claimed that the causal objection does not apply to mathematics because mathematics is not about objects (particulars). Rather it is about universals (usually higher-order relations of some sort) or about patterns or structures.[8] But the question still arises as to whether these entities have spatio-temporal locations and causal powers. If they do not, then it remains a puzzle as to how we can have knowledge of them. If they do, then they are not platonic entities as I conceive them. Given a clear formulation of the nature of non-particular or incomplete entities (not easy to come by), I claim that the arguments I employ against platonic particulars will also apply *mutatis mutandis* to platonic non-particulars, but I shall make no further argument for this.

The distinction between the concrete and the abstract on the basis of their causal powers or lack of them has been questioned (Resnik 1991). But if they cannot be distinguished on the basis of their causal powers or lack of them, this can only be because all objects have causal powers or none do. If the former, then the platonism that concerns me is false because there are no acausal objects. The latter entails radical eliminativism with respect to causation. My arguments are premised on the assumption that such elim-inativism is false. And most accounts of platonism explicitly or implicitly endorse the assumption that concrete objects have causal powers.

I assume that the platonist claim is a substantive one. My objections are not to be avoided by claiming that platonic objects lack physical causality, but can be known by means of some sort of non-physical causality. Entities

---

[7] According to Frege (1960), functions and concepts are not objects (p. 32), rather they are incomplete or 'unsaturated' entities (p. 153).

[8] See Bigelow (1988) & (1990) for an account of mathematical entities as universals. Tieszen (1989) characterises structuralism as the view that 'mathematics is really not concerned with individual mathematical objects at all but rather only with certain *structures*', (p. 19, his emphasis). See Shapiro (1997) for a detailed defence of mathematical structuralism.

with 'non-physical' powers are not platonic entities as I have defined them. However, some of my arguments, particularly in my discussion of the claim that we can have platonic knowledge by intuition (Chapter 9), could be employed to support the claim that we have no good reason to believe that there are any non-physical processes by which we can acquire knowledge.

My discussion of platonism in this chapter should be seen as scene setting. Many of the claims concerning platonism that have been raised are not essential to the platonist doctrine that is the target of my thesis. The minimal platonist claim with which I take issue may be stated as follows:

(MP)   Human beings can and do have knowledge of mind-independent objects that (unlike some other objects) are wholly lacking in the power to causally interact with human beings.

Finally, all anti-platonists must concede that our language is saturated with terms which may be construed platonistically, and that it is impossible to avoid using such language. Although I attempt to avoid language that entails a blatant commitment to platonism, for the sake of clarity, elegance, and the comfort of the reader, I shall not strive officiously to avoid all 'apparently platonic' talk. Use of such talk does not contradict the claim that there is an epistemological problem for platonism arising from the acausality of platonic objects.

1.7. OUTLINE OF MY PROJECT

My project is to examine and elaborate the epistemological problems for platonism that arise from the acausality of platonic objects. I defend three major claims:

(1)   A necessary condition for knowledge is the existence of a causal connection between fact and belief and there can be no such connection for the facts required by platonism.

(2)   Even if (1) is not accepted as a condition for knowledge of all facts, knowledge of the existence objects does require that such objects play a causal role in the belief in their existence.

(3)   Even if (2) is not accepted as a condition for all existential
           knowledge, the burden of proof is still on platonists to provide
           an epistemology for platonic knowledge.

I first defend (1) against some general objections including the claims
that facts cannot be causes or enter into causal relations (Chapter 2) and that
the etiology of a belief is, or may be, irrelevant to its status as an item of
knowledge (Chapter 3).

Next, I argue for the plausibility of (1) in Chapter 4, and then defend it
against purported counterexamples (Chapter 5). Although (1) remains de-
fensible, its plausibility is somewhat dented by these criticisms.

In Chapter 6, I consider relevant conditions on knowledge suggested by
alternative theories of knowledge and show that those that are plausible also
rule out platonic knowledge. I turn to the weaker claim (2) and argue for its
plausibility and defend it against purported counterexamples in Chapter 7.

In Chapter 8, I argue that even if we do not have an account of knowledge that rules out the possibility of knowledge of platonic objects, there is
still a reasonable requirement on platonists to give some account of how it is
that human beings have such knowledge.

Finally, in Chapters 9 to 12, I criticise the four major accounts of platonist knowledge to have been proposed, namely, knowledge by intuition,
apriorism, 'indispensable-postulate'-ism, and 'plenitudinous' platonism.

In two appendices, I defend claims that are not quite so central to my
thesis and which are assumed in the main body of the text.

CHAPTER 2

BELIEFS AND FACTS

2.1. WHAT ARE BELIEFS?

I do not intend to offer a full analysis of belief, but only to make explicit the basic assumptions about beliefs that underlie the arguments I employ. These assumptions should be uncontroversial. In particular, they are compatible with any current theories of mind except for eliminativism and perhaps naïve behaviourism. It is widely accepted that behaviourism, whether naïve or sophisticated, has failed, so I shall not argue against it here. I discuss belief eliminativism briefly in Section 2.3.

Beliefs are mental states. When Sally believes that her cat is lost, it is in virtue of her being in a certain mental state, which persists just so long as she so believes. That mental state just is her belief that her cat is lost. Acquiring a belief involves a mental change in Sally as does losing or discarding a belief.[1] Beliefs do not have to be conscious mental states. We can have a belief without being aware of it. I have believed for many years that Rabat is the capital of Morocco, but have seldom consciously entertained the matter during that time. Thus, I reject the Humean theory that a belief is 'a lively idea related to or associated with a present impression' (1739-40/1978, p. 96). However, we must beware of ascribing a plethora of occurrent, subconscious beliefs. If I am asked if I believe that the planet Mercury is heavier than Madonna's left knee-cap, then I am likely to reply that I do

---

[1] What one takes as constituting these states and changes will depend on one's theory of mind. Those with a taste for austere ontology may regard all belief ascriptions as predications of properties or relations to persons. The claim that people have beliefs is then no more than the claim that people believe.

believe that, but it is doubtful that I believed it before the question was asked. I acquire the belief as a result of my considering the question. To suppose that I had the belief all along would mean that I have an extremely large (probably infinite) number of subconscious beliefs. Only a dualist theory of mind could accommodate such a claim (assuming that if the mind is some sort of physical system, then it has only a finite amount of belief-storage space). On the other hand, it may be that we do have beliefs of which we have never been conscious. But this is an empirical matter, one to be decided by scientific investigation. Obviously introspection cannot be a reliable guide on the question of which unconscious mental states we have.

Beliefs are propositional attitudes. They belong, along with desires, doubts, fears, hopes, and so on, to that class of mental states which have *contents*. Propositional attitudes may be recognised by the 'that'-clauses embedded in sentences that ascribe them. For example, 'He believes *that* Elvis Presley is alive', 'She doubts *that* the world will end next week', and 'They hope *that* the next prime minister will be a woman'. The contents of propositional attitudes are the propositions picked out by the 'that'-clauses. But what this amounts to is highly problematic. Two sorts of problems arise. First, there are ontological issues concerning the nature of propositions. What sort of entities are they? There is a long tradition of regarding them as platonic entities. Alternative accounts of the contents of propositional attitudes designed to avoid appeal to platonic entities are diverse, controversial, and incomplete. Secondly, there are problems concerning the determinacy of mental contents. Do beliefs have a determinate content? What is it that determines content? A solution to these problems is beyond the scope of this book. Aspects of some of them are addressed in this chapter. Others are referred to in later chapters when relevant to a specific issue. Otherwise, they must regrettably be set to one side.

However, I shall continue to avail myself of proposition-talk.[2] Such talk is useful and convenient, and, given the current state of play, seems unavoidable if I wish to avoid committing myself to any particular ontological or psychosemantic theory. Proposition-talk is particularly useful in the way it unifies different aspects of belief-ascription. If Mereana believes that the house is big and asserts her belief by means of the sentence 'He nui te whare', then we can say that the very same proposition (namely, that the house is big) is the content of her belief and is the meaning of her sentence and is the truth-bearer of that which makes her belief true.

---

[2] My attitude to propositions is similar to that of Goldman (1986, pp. 13-17).

There are, roughly, three possible outcomes on the proposition-talk issue. Either such talk is completely inappropriate and must ultimately be discarded along with all talk of beliefs and other propositional attitudes. I touch on this eliminative possibility in Section 2.3. Or such talk (or a refinement of it) is appropriate, but it can be shown that it does not entail the existence of platonic entities.[3] Or it is irreducibly about platonic entities. If the last, so much the worse for my project. Or is it? If our best theory about the contents of mental states is a platonic theory, then it will still need to face the epistemological problems that I raise for such theories. Newton's theories may have been the best available, but that could not be given as an excuse for ignoring such problems as the precession of the perihelion of Mercury. My project is to present the causal case against platonism. Many of the difficulties raised by that case would still remain even if we were to conclude that platonism is our best available theory. In particular, we would still require an epistemology for acausal entities.

The term 'belief' may be used ambiguously. Sometimes it refers to a mental state, sometimes to the content of such a state. Where it is necessary to avoid ambiguity, I shall employ the terms 'belief-state' and 'belief-content', respectively.

2.2. STRONG AND WEAK CAUSAL THEORIES

I wish to argue for the claim that a necessary condition for knowledge is the existence of a causal connection between fact and belief and that there can be no such connection for the facts required by platonism. The first part of this claim may be expressed as:

(SC)   S knows that p only if the fact that p is causally connected to S's belief that p.

A theory that requires a causal link between a fact and a belief in that fact may be called a 'strong' causal theory of knowledge, in contrast to a 'weak' causal theory which simply includes a condition to the effect that:

---

[3] For example, if we cannot analyse the semantic value of beliefs without recourse to propositions and the semantic value of a belief makes a difference to its causal powers, then this would suggest that propositions are not acausal.

(WC)   S knows that p only if S's belief that p was caused in an 'appropriate' way.[4]

Bob Hale has claimed that only a strong causal theory can threaten platonic knowledge. I examine that claim in Chapter 7.

What constitutes an 'appropriate' way for a belief to be caused will depend on the details of the particular theory. For example, one version of reliabilism has it that the appropriate way is for the belief to be caused by a reliable process (e.g. Goldman 1986, p. 43). (SC). itself is a variety of (WC), since being causally connected to the fact that p is a particular way in which the belief that p may be caused. At least, this is so for the interpretation of (SC) that I develop and defend. In this and the following chapter, I examine a number of objections that threaten (SC). Some of those objections are aimed at (WC) in general. Sometimes it will be more convenient to defend (SC) by defending the wider claim (WC). It will also be useful to have defended (WC) against those objections when some alternative theories of knowledge are examined in Chapter 6. In Chapter 5, I examine objections specifically aimed at (SC).

2.3. BELIEF ELIMINATIVISM

It has been suggested (most notably by Churchland 1981) that we do not have beliefs; that beliefs do not exist, just as unicorns and phlogiston do not exist. If we do not have beliefs, then no one can meet the conditions of (SC) or (WC). There is a considerable literature on belief eliminativism, but I shall not discuss it here. The case for eliminativism is interesting and sophisticated, but my view is that, although the future discoveries of science may revise and extend our notion of belief, they are unlikely to show that we have no beliefs.

Suppose I am wrong and the belief eliminativists carry the day. Then either knowledge is eliminated along with belief or knowledge is possible without belief. If the former, then, since we cannot have knowledge of anything, we cannot have knowledge of platonic objects and epistemological platonism is false. If the latter, there is nothing in the eliminativist programme that suggests that knowledge is possible without the knower being in *some* appropriate state. In the absence of any detail as to what sort

---

[4] The weak/strong distinction is taken from Hale (1987, p. 93). The formulation of (SC) is suggested by Goldman (1967, p. 369).

of state that might be, I suggest that in a 'post-belief' era (SC) and (WC) would become something like:

(SC′)   S knows that p only if the fact that p is causally connected to S's appropriate state for knowing that p, and:

(WC′)   S knows that p only if S's appropriate state for knowing that p was caused in an 'appropriate' way.

As far as I can tell, causal objections to platonism would remain substantially the same.

2.4. TWO TRADITIONAL CONDITIONS FOR KNOWLEDGE

The strong causal condition on knowledge states that:

(SC)   S knows that p only if the fact that p is causally connected to S's belief that p.

(SC) entails that knowledge requires both truth and belief. This accords with what has become known as the traditional analysis of knowledge:

(TK)   S knows (at t) that p if and only if:
      (a) p is true
      (b) S believes (at t) that p
      (c) S is justified (at t) in believing that p. (Gettier 1963, p. 121)

I discuss and endorse the claim that conditions (a) and (b) are necessary conditions for knowledge in Appendix I.

Some theories concerning truth, especially those that arise from a concern that the notion of truth itself is troublesome, may raise difficulties for condition (a). But (a) should be read as requiring truth in a minimalist sense. The condition may be expressed simply as:

(TC)   S knows (at t) that p only if p.

Acceptance of the truth condition in this form does not assume any particular theory of truth, not even a minimalist theory, since (TC) does not mention truth at all.

2.5. AN ANALYSIS OF FACTS AS CAUSES

The strong causal condition on knowledge:

(SC)   S knows that p only if the fact that p is causally connected to S's belief that p.

has been criticised because, although (SC) ascribes causal powers to facts, facts are not entities which can enter into causal relations (Steiner 1975, pp. 111-13). One ground for this criticism is that facts are such dubious entities that it is doubtful that they exist. Another is that if they do exist, then they are abstract entities and so lack all causal powers. A third ground is that entities of some other type are the actual causal relata. If this criticism is sound then, since facts do not cause anything, either (SC) is false or we do not have any knowledge.

I present an analysis of causal claims of the form:

(F)   The fact that p caused S to believe that p

for certain propositions.[5] If my analysis is accepted, then I shall have demonstrated that the condition in (SC) can be met in at least some cases. My immediate task is to show that (SC) does not rule out the possibility of *any* knowledge. In a subsequent chapter, I address the question as to whether (SC) is a necessary condition for *all* knowledge.

I intend that my analysis be compatible with any theory of causation, at least with any reasonable or plausible theory of causation. There are two provisos. First, I reject, for obvious reasons, theories that are committed to the existence of platonic objects. Secondly, for any claim that A causes B, it need not be the case that A is a sufficient condition for B. In other words, it is not intended that A be the full cause of B. For example, a car accident may be caused by the inattention of the driver, the wetness of the road, a dog running in front of the car, and even the particular position of a power-pole. Each of the latter is a cause of the accident, none by itself is *the* cause, and even together they probably do not constitute the full cause. John Mackie (1974) captures the distinction between a cause and the full cause with his notion of a cause as an *inus* condition. An *inus* condition is 'an insufficient but non-redundant part of an unnecessary but sufficient condition' (p. 62).

---

[5] For an alternative, and more detailed argument for facts as causal relata, see Mellor (1995).

I start with claims of form (F) where *p* is an *atomic* proposition. An atomic proposition consists of an *n*-place predicate and *n* singular terms. For example:

(1)   Mark is tired (at time *t*).[6]

This yields the causal claim of form (F):

(2)   The fact that Mark is tired caused S to believe that Mark is tired.

The same claim can be expressed in more natural English as:

(2′)   S believes that Mark is tired because Mark is tired

where context usually indicates that 'because' is being used in a causal sense. But the same claim can also be expressed by:

(2″)   Mark's being tired caused S to believe that Mark is tired.

All claims of type (F) where *p* is an atomic proposition can be reformulated in a similar way. In each case the proposition is formed into a noun phrase by the process of *nominalisation*. Here are some further examples:

(3)   'Mark tripped' becomes 'Mark's tripping',

(4)   'The mayor departed' becomes 'The mayor's departing' or 'The mayor's departure',[7]

(5)   'The princess kissed the frog' becomes 'The princess's kissing the frog',

and so on. Negations of atomic propositions can be reformulated in a similar way. For example:

---

[6] Assume that all of the following propositions are relativised to a specific time or time interval.

[7] Perhaps these are not strictly equivalent. English allows us to make subtle distinctions. 'The mayor departed' corresponds to 'The departure of the mayor', while 'The mayor was departing' corresponds to 'The mayor's departing'.

(6)   'Sue is not laughing' becomes 'Sue's not laughing',

(7)   'The match did not light' becomes 'The match's not lighting' or
           'The failure of the match to light'.

(8)   'Sue has no money' becomes 'Sue's not having any money' or
           'Sue's lack of money'.

For anyone who finds statements such as:

(9)   The mayor's departure caused S to believe that the mayor
           departed

and

(10)   The failure of the match to light caused S to believe that the
           match did not light,

unobjectionable, little more need be said. If statements such as (9) can be
true, then (SC) can be met, at least for cases where p is an atomic proposition
or the negation of an atomic proposition.

2.6. FACTS AND ONTOLOGICAL COMMITMENT

One concern which may arise with respect to this analysis of causal claims is
that it carries with it unacceptable ontological commitments. This is not the
case. But in order to show this, various possible ontologies must be
considered. After all, what may be ontological poison to one person may be
ontological meat to another. For my purposes, the only ontological poison is
the existence of entities that lack causal powers. Within that constraint, I
shall demonstrate that (SC) is compatible with ontologies from the lush to
the arid.

First of all (SC) does not commit us to the existence of facts. Claims like
(2) can be regarded simply as *façons de parler* for claims like (2''), (9) and
(10). But what do those claims commit us to? A natural interpretation is that
they cite events and conditions as the causes of S's beliefs, and there are

good arguments for regarding events and conditions as the basic causal relata.[8]

What are events and conditions? The nominalisations discussed above give strong support to the notion that events and conditions are the descripta of certain true propositions. To be precise, they are the descripta of true atomic propositions and true negated atomic propositions.

It may be useful to retain the term 'fact' rather than dismiss it as a mere *façon de parler*. We can simply take the class of facts[9] as being identical to the class of events and conditions.[10] It should be noted that 'fact' is a notoriously ambiguous term. In one sense, it can simply mean 'true proposition'. In another sense, it refers to quasi-linguistic abstract entities that are purported to be the denotations of true propositions (Barwise & Perry 1983 and Taylor 1985, ch. 2). However, neither of these is the sense in which the term is employed in causal claims like (F). From such claims, we can simply read off a genuine causal statement about events or conditions.

A common conception of facts *qua* events and conditions is as *structured complexes* (Menzies 1989, p. 69; also Kim 1975). According to this conception, a number of different kinds of entity put together in a certain way constitute a fact. For example, the condition which is Mark's being sleepy at *t* consists in Mark (a physical object) exemplifying the (physical) property of sleepiness at time *t* (or during the time interval *t*), where exemplification is a primitive and unanalysable relation holding among the constituents of that condition. On the other hand, if Mark was not sleepy at *t*, we would have the condition of Mark's not being sleepy at *t*, which would consist in Mark not exemplifying the property of sleepiness at time *t*, where non-exemplification is a different (primitive and unanalysable) relation from that of exemplification (Menzies 1989, p. 69). One advantage of construing facts in this way is that it allows for a straightforward account of identity conditions for facts.

But does this construal commit us to the existence of facts over and above the existence of the objects, properties, and times that are their constituents? Certainly facts depend for their existence on those constituents, but do they supervene on them? It would seem not. Mark, sleepiness and time *t* could all exist and yet it not be the case that Mark is sleepy. So this

---

[8] See, for example, Menzies (1989), except that what I call conditions, he calls states of affairs.

[9] Strictly speaking, the class of facts associated with atomic propositions and negations of atomic propositions.

[10] See, for example, Menzies (1989), except that what I call *facts*, he calls *situations*. It is clear from the literature that, whatever terminology one adopts, it will clash with someone else's. Stipulation is the order of the day.

construal takes facts to be as much real features of the world as physical objects, though categorically different from them. Whether or not this is acceptable will be a matter of ontological 'taste'. Below, I discuss alternative construals that avoid this ontological inflation.

Another (perhaps more serious) objection to construing facts in this way is that it commits us to the existence of properties and relations. Are not properties and relations universals? And are not universals abstract entities, and hence acausal? The answers to both of these questions need not be in the affirmative. My project is to examine the propriety of positing acausal objects. The property of sleepiness (if there be such a thing) can hardly be devoid of causal powers since it makes a difference to Mark's causality whenever he possesses that property. However, many realists about universals claim that properties exist whether or not they are actually being exemplified. Construed in that way, universals must be eschewed by the causalist since it implies that there are universals that exist outside the causal order. But there are other ways of construing universals. Armstrong (1978b) claims that universals are *physical* entities that cannot exist unless they are exemplified by physical individuals. So one way we could construe facts as structured complexes without running foul of causal objections is along Armstrongian lines.

But we may balk at the notion of universals as entities that wholly exist in more than one place at the same time, an essential feature of Armstrong's account. An alternative way of viewing properties is as *property-instances* or *tropes* (Williams 1966 and Campbell 1990). On this view, Mark's sleepiness at *t* is something that only Mark can have. It is similar, but in no way identical, to other sleepinesses, including Mark's sleepinesses at other times. Accordingly, it is Mark's having the trope that is *his* sleepiness at *t* that may be the cause of some effect (e.g. S's having a certain belief).

This account appears to have difficulties with 'negative' facts. Typically, the trope theorist does not suppose that when Mark is not sleepy it is in virtue of his exemplifying a different property, say non-sleepiness. And the relation of non-exemplification is redundant to this account. For the tropist to say that Mark is not sleepy is simply to say that his property of sleepiness does not exist. Mark cannot bear a relation to something that does not exist. The problem is that, although we may accept that the trope which is Mark's sleepiness has the causal power to give rise to a belief, if Mark is not sleepy what is it that has the causal power to give rise to the belief that Mark is not sleepy? The tropist will have to argue that in such a case Mark will have other properties, some conjunction of which will be incompatible with his being sleepy, and that it is this conjunction of properties that has the power to cause the belief.

The more austere nominalists will wish to eschew properties and relations altogether and claim that only particulars exist. But they will not wish to say that it is Mark *simpliciter* that can be the cause of certain effects. After all, Mark need not have been sleepy. If he were not sleepy then his causal powers would have been different. So what is it about Mark that makes the difference if not his having or exemplifying a certain property? The austere nominalist can claim that individuals not only exist, but can and do exist in different ways (Lewis 1992, p. 218). When Mark is sleepy it is in virtue of his existing in a sleepy way and when he is not sleepy it is in virtue of his existing in a non-sleepy way. If he exists in the former way then he will have certain causal powers, and if he exists in the latter way he will differ in his causal powers. Any request for a deeper analysis of Mark's state is rejected as arising from a 'pseudo-problem', unless it is a request for a scientific investigation into the underlying mechanisms of his state (and similar states).

Each of the positions I have examined is problematic; each is controversial. If that were not so, then we would not be faced with this plethora of rival positions. But the problems each one faces are wider than that of accounting for causal claims that take events and conditions as the basic causal relata. It is up to the proponents of each position to deal with the problems facing their position. Those problems aside, each position can deal with the causal claims.

For my purposes, the claim that the fact that p can cause S to believe that p, where p is an atomic proposition or the negation of an atomic proposition, can be regarded as primitive. What I have shown is that if an analysis is sought, then it can be provided without resort to dubious entities; in particular, without resort to platonic entities.

2.7. COMPLEX FACTS

I have completed my task of showing that the condition required by (SC) can be met in at least some cases, namely, those where the belief-content is an atomic proposition or the negation of an atomic proposition. I shall indicate briefly how my analysis may be extended to cover knowledge of more complex facts.

In the case of conjunctive facts, the claim that the fact that $p \& q$ caused S to believe that $p \& q$ is equivalent to the claim that both the fact that p and the fact that q caused S to believe that $p \& q$.

In the case of disjunctive facts, the claim that the fact that $p \vee q$ caused S to believe that believe that $p \vee q$ is equivalent to the claim that either the fact that p or the fact that q (or both) caused S to believe that $p \vee q$.

Existential facts, such as the fact that there exists an F, are covered by an extension of (SC), which I argue for in Section 5.4. I do not claim that such facts can be direct causes of beliefs, or of anything else for that matter. Unrestricted universal facts are more problematic and are discussed in Section 5.6. Once again, I shall argue that although they are not the direct causes of our beliefs in them, knowledge of them can be accommodated by an extension of (SC).

Even more problematic are modal, subjunctive, and normative facts. The truth conditions for statements of such facts are the subject of much current debate. For this reason, I shall set them aside. On some theories, such facts are reducible to, or supervenient on, facts of the kinds discussed above. If such a theory posits platonic entities, then it will face the epistemological problems that I raise for such theories. If not, then there is no problem so far as my project is concerned. Theories that deny that there are such facts are also not my concern. Theories that claim that such facts are primitive, irreducible, or unanalysable would stand in need, I suggest, of an account of how we can have knowledge of such facts. Each such account would then need to be assessed, piecemeal.

I conclude that the notion of 'fact' can be explicated to show that facts are not dubious entities and that they can be the causes of beliefs.

CHAPTER 3

BELIEFS AND CAUSES

3.1. FREE CHOICE AND CAUSALITY

In this chapter I explore various objections to the claim that for a belief to count as knowledge it must have been caused in some way. First, in this section, I consider the claim that many beliefs are uncaused and yet they can still constitute knowledge. If beliefs can be uncaused and if such uncaused beliefs can be items of knowledge, then both the strong causal condition:

(SC)   S knows that p only if the fact that p is causally connected to S's belief that p,

and the weaker causal condition:

(WC)   S knows that p only if S's belief that p was caused in an 'appropriate' way.

are false. I shall defend (WC) against this charge. Because (SC) is a restricted version of (WC), the arguments and counterarguments will apply *mutatis mutandis* to (SC).

Here is an argument that the causal requirement (WC) is false. If I freely choose to believe that p then, under exactly the same circumstances, I could have chosen not to believe that p (or perhaps even to believe something else). Given that the circumstances are the same in each case, any so-called causes operating in those circumstances can have played no role in bringing about my belief that p. So a freely chosen belief is an uncaused belief. But some freely chosen beliefs may be items of knowledge. Therefore, some items of knowledge, being uncaused, do not meet the causal requirement. Thus, the causal requirement is false.

It might be objected that when we freely choose to believe something, our belief has a cause. The cause of the belief is a free choice or decision. The free choice may be uncaused but the belief is not. But this is of little comfort to the causal theorist. Although the causal requirement leaves it open as to what is an 'appropriate' way for a belief to be caused, a theory would have little content as a *causal* theory if a belief's being directly or immediately caused by an uncaused free choice were deemed appropriate. So let us take 'uncaused beliefs' to include beliefs directly caused by an uncaused free choice.

Of course, determinists will reject the above argument as unsound. Determinists deny that there are any uncaused beliefs because they deny that there are any uncaused events.

Hard determinists will reject the premise that we can freely choose what to believe as well as the conclusion that we can have uncaused beliefs. They claim that we cannot make genuinely free choices. Compatibilists may accept the premise while denying the conclusion. They reject the notion that a free choice is an uncaused event, claiming that free choices are caused events like any others.

Only libertarians, who reject both hard determinism and compatibilism, would accept the conclusion that there may be uncaused beliefs. Are causal theories of knowledge then incompatible with libertarianism? It would be undesirable for causal theories to beg the question against libertarianism. But is there a sound argument for this supposed incompatibility? The argument would appear to be as follows:

(1)   We can make free choices that are uncaused events (libertarianism)
(2)   We can freely choose at least some of our beliefs
      therefore
(3)   We can have uncaused beliefs
      but
(4)   Uncaused beliefs may be items of knowledge
      therefore
(5)   The causal requirement (WC) is false.

All of the premises of the argument may be questioned. Consider premise (2), the claim that we can freely choose our beliefs. This claim is the position of belief-voluntarism. In Appendix II, I explore belief-voluntarism in detail. In this section, I assess it briefly.

If premise (2) means that we can *directly* choose our beliefs, then it is false. Believing is not something we have control over in the way that we may have control over our actions. As Hume puts it:

> We may, therefore, conclude, that belief consists merely in a certain feeling or sentiment; in something, that depends not on the will, but must arise from certain determinate causes and principles, of which we are not masters (1739-40/1978, p. 624).

On the other hand, if (2) means only that we can *indirectly* choose our beliefs, then the argument is invalid. Suppose it is observed that people exposed to a particular environment tend to form a particular belief. If I decided that I should like to have that belief, could I not then acquire it by choosing to expose myself to that environment? This is what Pascal had in mind when he suggested that by adopting a religious life one could come to believe sincerely in God. He suggests that '... taking holy water, having masses said, and so on ... will make you believe quite naturally' (1670/1966, p. 152). The important thing to note about this method of belief acquisition is that rather than giving support to the view that beliefs may be uncaused, it relies on the fact that beliefs are causally influenced by our environment or way of life. What is chosen is a course of action. It is then up to external causes to form the desired belief.

To sum up, either premise (2) is false because we cannot directly choose our beliefs, or it is true in virtue of our being able to indirectly choose (some of) our beliefs. If the latter, then sub-conclusion (3) that we can have uncaused beliefs does not follow, since any process by which we indirectly choose to believe turns out to be a process by which we choose to act so as to expose ourselves to causal influences that may give rise to a particular belief. Such processes are compatible with a causal requirement on knowledge.

3.2. CAN UNCAUSED BELIEFS BE KNOWLEDGE?

Suppose I am wrong about belief-voluntarism. Suppose I am wrong because we do have the ability to choose our beliefs directly (in a libertarian sense). There is another objection to the soundness of the incompatibility argument. The sub-conclusion (3) that we can have uncaused beliefs may be accepted, but claim (4), that uncaused beliefs may be items of knowledge, rejected. The objection to (4) is that uncaused beliefs cannot meet plausible criteria for being knowledge. Would not an uncaused belief be an uncontrolled or random belief, and is not an uncontrolled or random belief just the sort of belief that would not count as an item of knowledge?

There are two possible anti-causalist responses. The first is to claim that a belief counts as knowledge because it meets certain criteria that are independent of the etiology of the belief. For example, on a coherence theory of knowledge, the criterion might be that the belief should cohere with the believer's other beliefs. Beliefs acquired by a random or uncontrolled process could meet such a criterion. But this response introduces a quite different argument against the causal requirement, one that does not depend on our having freewill. I discuss that argument in Section 3.5.

The second anti-causalist response is to agree that random or uncontrolled beliefs could not count as knowledge but to deny that uncaused beliefs are necessarily random or uncontrolled. There are two (possibly overlapping) arguments for this position.

The first rests on the claim that free choices may be under the control of a rational agent. When rational agents survey the available evidence they will freely choose a belief that is compatible with and supported by that evidence. If that belief is also true then it will be a potential item of knowledge. This approach stems from the agent-causalist theory of freewill according to which free choices are caused by agents but agents themselves are not caused. In other words, agents have the ability to start new causal chains. Strictly speaking, this position may be compatible with the requirement that beliefs be caused in an 'appropriate' way, since being caused by a rational agent may be deemed an 'appropriate' way for a belief to be caused. The causal chain in such cases is very short. But such minimal causal chains are incompatible with substantive causal theories that clearly appeal to processes involving much longer causal chains, often chains which extend back to causes external to the knower.

The second argument rests on the claim that an uncaused belief need not be totally uncaused. When we say that the explosion was caused by the striking of a match this does not rule out the assertion that the explosion was caused by the leaking of gas. The striking of the match and the leaking of the gas are partial causes which, combined with other events and conditions, make up the total cause of the explosion. Similarly, a number of causes combined with a free choice may give rise to a belief. For example, I read in the newspaper that rain is forecast and I hear rain-like noises coming from the roof. I choose to believe that it is raining. My belief is not (entirely) random or uncontrolled because it was partially caused by what I read and heard. (Those who believe that the cause must be in some sense sufficient for its effect will reject such a notion, but the aim here is to be as generous as possible to the libertarian view.)

From these two arguments we can extract three positions, which oppose the view that freely chosen beliefs are random or uncontrolled:

(F1)   A freely chosen belief is completely under the control of a rational agent.

(F2)   A freely chosen belief is partially caused and has an uncaused element.

(F3)   A freely chosen belief is partially caused and partially under the control of a rational agent.

But (F2) and (F3) are compatible with the causal requirement. Causal theories need not insist that the *total* cause of a belief be 'appropriate' and many do not. In particular, any plausible version of (SC) will be concerned only with a partial cause, since it will not be supposed that the fact that p will (necessarily) be the total cause of the belief that p. So if position (F2) or (F3) is adopted and (4) is read as:

(4′)   *Partially* uncaused beliefs may be items of knowledge

then the argument is not valid since the falsity of the causal requirement does not follow from (4′).

On the other hand if (4) is read as:

(4″)   *Totally* uncaused beliefs may be items of knowledge

then neither position (F2) nor (F3) will be compatible with it, because both those positions allow that freely chosen beliefs are partially externally caused.

Of course, each of those positions may have a limiting case where the external cause is nil. For (F2), the limiting case would be a freely chosen belief that is entirely random. Since we are considering the position of those who accept the premise that random or uncontrolled beliefs could not be items of knowledge, (4″) would be false in such cases and the rejection of the causal requirement not justified. For (F3), the limiting case would be a freely chosen belief that is completely under the control of a rational agent, in other words, identical to position (F1).

I conclude that the only way an uncaused belief could be an item of knowledge would be if that belief were a freely chosen belief that is completely under the control of a rational agent. Is it plausible that a freely chosen belief could be completely under the control of a rational agent? And is it plausible that such a belief would qualify as an item of knowledge? Certainly not if it involved the rational agent's choice being made without any regard to her perceptions, memories, inferences or other mental states. If that were the case, then the belief could only be true by chance and one of the more important desiderata (if not the most important) for any theory of knowledge is that it rule out such lucky beliefs as genuine items of knowledge. But if the agent *is* influenced by her mental states, will not those mental states then have a causal role, so that the belief will not be totally uncaused after all? Not on one view of agent causalism. This is the view that the agent may have regard to the *contents* of her perceptions, memories and inferences (i.e. the evidence considered as propositions) rather than being causally influenced by those perceptions and memories as mental events or states. So a totally uncaused belief may be one where the rational agent holds the evidence, as it were, at 'arms-length'. The rational agent reflects on the evidence that she brings before her mind. The contrast is between, for example, on the one hand, hearing scratching noises behind the wall and coming automatically to believe that there is a mouse there and, on the other hand, reflecting on the fact that one has heard those sounds and deciding to believe there is a mouse there. Another example of a freely chosen belief under the control of a rational agent might be that of a philosophy student who considers all the evidence for and against the respective positions of mind-body dualism and materialism. Both positions seem plausible, but he chooses to believe one of them.

These scenarios are reminiscent of Kant's account of the will of a rational being (Paton 1948, ch. 3). In them I refer to the crucial relation between agent and evidence as that of 'reflecting on', 'having regard to', or 'considering'. It may be objected that these (albeit metaphorical) terms refer to relations which are essentially causal relations. I share this worry. But we can set this worry aside, as my aim is to provide as sympathetic an account of agent causalism as I can muster. I am a belief-involuntarist, so I do not believe that such scenarios are possible for human believers. But we are supposing that I am wrong about belief-voluntarism. Given that it is possible for freely chosen beliefs to be completely under the control of a rational agent, the question now is whether or not such beliefs could count as knowledge. I do not see why not. If the evidence is adequate and is carefully considered, and the belief is true, then it is plausible that the agent could not only believe, but also know. If that is so, then the causal requirement on knowledge would be false. But rather than abandon the causal requirement, I suggest an amendment to it.

3.3. AN AMENDMENT TO THE CAUSAL CONDITION

Consider the evidence that the rational agent is surveying. According to the scenario, the agent is reflecting on evidence that she brings before her mind. What she is reflecting on must be the contents of mental states. These mental states must be either caused or uncaused. It is implausible that an agent could know as a result of choosing to believe the content of an uncaused mental state. Such a mental state will have occurred at random or possibly the agent will have freely conjured it up by an act of imagination. Either way, to choose to believe the content of such a mental state cannot result in the acquisition of knowledge. On the other hand, if the mental representation of the evidence has been caused, then it may, or may not, have been caused in an epistemically 'respectable' manner. If it has its origins in a dream or a hallucination or the testimony of a renowned deceiver, then to believe its content will not result in the acquisition of knowledge.

I suggest the following amended versions of the causal condition on knowledge:

(WC'')   S knows that p only if either:
   (a) S's belief that p is caused (or partially caused) in an 'appropriate' way, or
   (b) S's mental representation of the evidence for p is caused (or partially caused) in an 'appropriate' way.

(SC'')   S knows that p only if either:
   (a) the fact that p is causally connected to S's belief that p, or
   (b) the fact that p is causally connected to S's mental representation of the evidence for p.

(WC'') and (SC'') are compatible with the intuitions, evidence and arguments that causalists typically appeal to in support of their theories. I suggest that all theories of knowledge with a causal requirement be amended to include clause (b), or that they be read as if they included clause (b). I suppose the latter to be the case in subsequent chapters.

Libertarianism is no threat to suitably amended causal conditions on knowledge, either because we cannot freely choose our beliefs, or because if freely chosen beliefs are to qualify as knowledge they must also be partially caused or the evidence for them must be caused or partially caused.

3.4. NON-CAUSAL CRITERIA FOR KNOWLEDGE

Another objection to causal conditions on knowledge is that the way in which a belief is caused (if it is caused) is irrelevant to whether or not it qualifies as an item of knowledge. First I demonstrate that purported counter-examples to (WC), and hence to (SC), based on this objection are unsuccessful. Then I argue against accounts of knowledge that reject the relevance of belief etiology.[1]

A typical example, which purports to show that the epistemological status of a belief is independent of the causal antecedents of that belief, is as follows. A mother believes her son to be innocent of the crime of which he has been accused, although the available evidence points strongly to his guilt. She has no evidence for her belief. The belief arises from her consuming love for her son. But he is in fact innocent. Later, at the trial, the defence lawyer presents new evidence and argues convincingly that the son is innocent. The mother follows this argument and understands that it establishes her son's innocence. Before the trial she believes that her son is innocent but does not know that he is. After the trial she both knows and believes that he is innocent. Yet, the 'inappropriate' cause of her belief remains the same.

My reply to such examples depends on distinguishing earlier states of a belief from later states of that belief. After the trial the mother's belief is reinforced or sustained by the lawyer's presentation. It is this new causal process that legitimates her knowledge. Later states of her belief do not just depend on maternal love; even if she lost that love she would still believe in his innocence.

Lehrer's gypsy-lawyer example attempts to avoid this response (Lehrer 1974, pp. 124-25). A lawyer has a client who is accused of an horrific crime of which she is innocent. This client has previously committed similar crimes and there is overwhelming evidence that she has also committed this one. Even the lawyer believes she is guilty. But the lawyer is a gypsy and when he 'reads the cards' they tell him that his client is innocent. His

---

[1] My arguments are developed from Kitcher (1984, pp. 13-17).

unshakeable belief in the cards leads him to reconsider the evidence and he discovers a complicated but correct line of reasoning which conclusively demonstrates that his client did not commit the crime. Such is the subtlety of the reasoning and the influence of the client's previous criminal record that no-one else is convinced. Indeed, if it were not for his reading of the cards, the gypsy lawyer would also be swayed by those factors. Lehrer claims that the lawyer knows, on the basis of the evidence that he has examined, that his client is innocent. However, his belief that she is innocent is in no way supported or reinforced by that evidence. That belief cannot have been strengthened by the evidence, because he is already completely convinced by the cards. And if he lost his faith in the cards then, swayed by those other factors, he would no longer be convinced by the evidence. So his faith in the cards, which is the sole cause of his belief, cannot be relevant to that belief's status as knowledge.

Lehrer equivocates on the role of the evidence and the complicated reasoning. If the lawyer's knowledge of his client's innocence is based on that evidence and reasoning, then his examination of the evidence and his reasoning process must play a role in supporting or sustaining his belief in her innocence. Otherwise the lawyer would be in the position of someone who understands an argument but does not psychologically connect the argument with its conclusion. Such a person cannot be said to have knowledge even if he believes the conclusion (assuming he has no other appropriate grounds for his belief). No matter how convinced the lawyer already is, his understanding and appreciation of the evidence must give added support to that conviction. The fact that he would lose his belief if he lost his faith in the cards does not mean that his present belief is not sustained in part by his understanding of the evidence.

An analogy should help to make this clear. Suppose a tabletop is securely supported by three legs. A fourth leg is inserted flush beneath the tabletop. That leg now plays a causal role in supporting the top (cf. Goldman 1967, p. 362, fn. 7). It does so, even though the removal of the original three legs would result in the collapse of the table. On the other hand, the fourth leg could have been inserted so as not to bear any of the weight of the top. As Lehrer describes it, the lawyer's evidence and reasoning seem to be playing a role like that of the leg in this latter scenario. If so, the lawyer lacks knowledge. Otherwise, the evidence and reasoning are playing a role in sustaining his belief and (WC) is not refuted.

Part of Lehrer's confusion arises from his failing to distinguish between justifying the *content* of a belief and justifying the *having* of that belief.[2] I maintain that the former notion is incoherent. Whenever we speak of a belief's being justified, we only make sense if we mean that there are reasons why someone is, or would be, justified in holding that belief. Lehrer seems to think that there is a justification for the belief that the client is innocent and the lawyer is justified in believing his client innocent because he is in possession of that justification. What is that justification? It appears to be an argument from certain premises to the conclusion that the client is innocent. But argument is a matter of logic. Arguments entail or (more controversially) make probable their conclusions. They do not justify their conclusions. If that is what is meant by justification then the argument '*p* therefore *p*' would be a justification for any proposition *p* (cf. Miller 1987, pp. 346-51). What does make sense is to say that following an argument could justify someone's having a belief. But following an argument is a psychological event and it is the effect of that event which is crucial, whether that effect be the adoption of a belief or the sustaining of a belief. Having good evidence is not enough; it is what you do with it (or what it does to you) that determines whether or not you have knowledge.

3.5. NON-CAUSAL EPISTEMOLOGIES

Philip Kitcher (1984, pp. 13-14) divides theories of knowledge into two categories: *psychologistic* and *apsychologistic*. According to him, psychologistic theories are those that meet the requirement of (WC). They are theories that require that beliefs be caused in an appropriate way if they are to count as items of knowledge. He chooses this terminology because belief-producing processes 'will always contain, at their latter end, psychological events' (p. 13). But his terminology is doubly unfortunate. First, psychologism is also used to denote other positions, which, although similar in some respects, are not so philosophically respectable. For example, the theory that logic should be no more than a branch of psychology that describes the ways in which people think. This is the 'psychologism' to which Frege objected so strongly (1964, pp. 12-13). Secondly, epistemologies that lack a causal requirement may still have a psychological component. For example, Descartes' requirement that a belief

---

[2] See Musgrave (1989 pp. 332-33) for more on this distinction.

be clearly and distinctly conceived may be a psychological, but not a causal, requirement.

I shall argue that no epistemology that lacks (WC) can give a complete account of knowledge. I call such theories *non-causal*, but distinguish those with a psychological component from those without one.

3.6. APSYCHOLOGISTIC NON-CAUSAL THEORIES

Any plausible epistemology must have criteria for distinguishing knowledge from mere true belief. Apsychologistic non-causal theories have criteria concerned with the nature of the propositions believed and the logical relations between them, rather than criteria concerned with the nature of mental states and the psychological relations between those states. For example, such a theory may state that in order for a belief to count as an item of knowledge it must either be a self-evident belief or be a belief whose content is deducible from the contents of other self-evident beliefs. Alternatively, the criterion might be that the belief's content should cohere with all (or some) of the knower's other belief-contents.

Counter-examples to such theories are readily devised. Consider any true proposition which meets the proposed criteria (or which proponents of the theory consider meets the criteria). Imagine someone coming to believe in that proposition in some bizarre or disreputable way. For example, she comes to believe it because she reads it in a 'fortune-cookie' message, or because of an hallucination, or because it is implanted in her brain by aliens (who, themselves, have no idea of its truth). In such cases, she lacks knowledge.

Attempts can be made to patch up such theories, either by requiring that the knower have certain additional beliefs or by requiring that the original belief meet certain psychological requirements. If the latter, then the theory will no longer be apsychological. Suppose we require that the putative knower have certain additional beliefs. For example, a coherence theory may require that the knower also believe that the new belief coheres with all his other beliefs. But that belief, although true, could also be acquired in a bizarre manner, so the subject would still fail to know. Perhaps we should require that the subject *know* that the new belief coheres. Now we are caught in a regress. To count as knowledge, the additional belief will have to meet the criteria. Not only will it have to cohere with all the subject's other beliefs but he will have to believe that it does, and this extra belief that the additional belief coheres must, in turn, meet the criteria, and so on. Before we could know anything new, we would have to know an infinity of new things. It might be argued that this infinite regress is not vicious but benign, because we are capable of acquiring such an infinite set of beliefs. If we are, then it is also possible that the requisite infinite set be acquired in a bizarre manner; in which case, the subject will still fail to know. A purely apsychologistic theory of knowledge can never be adequate.

3.7. PSYCHOLOGISTIC NON-CAUSAL EPISTEMOLOGIES

Psychologistic but non-causal epistemologies are also inadequate. This is because true beliefs that meet the proposed criteria by having the required psychological attributes could still have been implanted in a bizarre manner. This problem plagues the Cartesian criterion of being 'clearly and distinctly conceived'. Why shouldn't an evil demon have implanted such a belief? Of course, the evil demon implants false beliefs, which, therefore, cannot be knowledge. But a frivolous demon could implant clear and distinct beliefs at random, some of which might happen to be true. I could clearly and distinctly believe that it is raining in Eketahuna and that belief may be true, but if my belief is caused by a frivolous demon or an hallucination then I do not know that it is raining in Eketahuna.

Cartesian attempts to avoid this problem often result in an equivocation on whether the criterion is psychological or not. Any argument to the effect that a clear and distinct belief is a belief that could not be false must consist in applying the criterion to the content of the belief rather than to the belief as a mental state. This sort of equivocation shows up clearly for a criterion like self-evidence. Applied to beliefs as mental states the criterion will be met by beliefs that the believer is psychologically unable to doubt. But there is no obvious reason why this psychological state should not apply to a belief with a false content. But, the Cartesian argument goes, self-evident beliefs are beliefs such as 'I think' or 'I exist'. Truths such as these are true if anyone contemplates them, so self-evident beliefs couldn't be false. But now self-evidence is being attributed to propositions, the contents of beliefs. The property of always being true under certain circumstances is not a psychological property so it is misleading to call it self-evidence. The next Cartesian move is to claim that propositions with that property will be believed by anyone who contemplates them, and that is why it is appropriate to call them self-evident. But this claim is perverse. There are no propositions that will inevitably give rise to belief, let alone the inability to doubt them, when contemplated.[3]

Either self-evidence is a psychological property of belief-states, in which case having the property will not guarantee truth and this in turn will allow the possibility of a self-evident belief's being fortuitously correct. Or self-evidence is an apsychological property of belief-contents which entails truth, in which case simply having such beliefs will not be sufficient to constitute having knowledge.

In this chapter I have demonstrated that various objections to (WC) are without force. I conclude that an epistemology is not plausible unless it contains the condition that for a belief to count as knowledge it must have been caused in an appropriate way.

---

[3] Musgrave (1993, p. 251) makes a similar point.

CHAPTER 4

THE CASE FOR A CAUSAL CONNECTION

4.1. DEFENDING THE STRONG CAUSAL CONDITION

So far, I have defended the claim that the etiology or causal history of a belief is crucial in determining whether or not that belief constitutes knowledge. Now I turn to the stronger claim that the fact that is believed must be causally connected to the belief in that fact in order that the belief be knowledge.[1] This is the strong causal condition:

> (SC)   S knows that *p* only if the fact that *p* is causally connected to S's belief that *p*.

If (SC) is accepted, then we cannot have knowledge of facts that involve entities lacking causal powers. In other words, we cannot have platonic knowledge.

(SC), or variations on it, has achieved prominence as a response to the Gettier problem. I shall argue that issues raised by the Gettier problem do provide grounds for accepting something like (SC). But first, I argue that something like (SC) underlies the concept of knowledge.

4.2. DISPUTING KNOWLEDGE CLAIMS

Paul Benacerraf (1973) points out that one way of disputing a knowledge claim is to argue that the knowledge claimant could not be, or have been, causally connected with the fact believed. Or as Benacerraf puts it:

---

[1] Much of this chapter is based on Cheyne 1997a.

[the claimant]'s four-dimensional space-time worm does not have the necessary (causal) contact with the grounds of truth of the proposition for [the
claimant] to be in possession of evidence adequate to support the inference (if
an inference was relevant). (p. 413)

One way of establishing the lack of such a causal connection is to provide a
plausible account of the belief's etiology, which causally explains the holding of the belief, but which does not involve the fact believed. Although this
may not conclusively establish the absence of a causal connection because
there is always the possibility of the belief's being causally over-determined
(as in Lehrer's case of the gypsy lawyer, Section 3.4), what is important is
that adopting such a strategy depends on the assumption that such a causal
connection is necessary for knowledge.

A good example of this strategy is to be found in Edward Gibbon's
account of the progress of the Christian religion (1776-88/1994, ch. 15).
Gibbon explains the persistence and increasingly widespread acceptance of
early Christian doctrine in terms of the Christians' intolerant zeal, austere
morality and discipline, their promise of immortality, and their claims to
miraculous powers. Although Gibbon states that these causes are only
secondary to the primary source of belief (namely, divine revelation), it
becomes apparent to the perceptive reader that his aim is to raise doubts
concerning this connection. He does not say this explicitly; it would have
been dangerous to do so in the political climate of the time. He relies on the
tacit assumption of the reader that if the having of those beliefs can be fully
explained without 'divine facts' playing a causal role, then it is reasonable to
suppose that the believers were not causally connected to those facts and,
consequently, cannot have had genuine knowledge of them. Furthermore, if
they lacked knowledge, then their having those beliefs in the way that they
did cannot play a role in warranting anyone else's beliefs in Christian
doctrine. It is only because Gibbon can rely on his readers to share this assumption, that he can make his point in this indirect way.

A similar strategy is apparent in the 'Strong Programme' approach to the
sociology of scientific knowledge. By arguing that social causes are the
determining factors in the formulation and acceptance of scientific theories,
adherents to this school imply that scientists do not obtain objective
knowledge of the physical world (Bloor 1976 and Brown 1989).

We do not have to accept that Gibbon's and Bloor's strategies are successful. We have only to see that such examples support Benacerraf's
contention that a causal connection underlies our conception of knowledge.

4.3. THE GETTIER PROBLEM

True belief is necessary for propositional knowledge. I make the case for this claim in Appendix I. But true belief is not sufficient for knowledge. I may come to believe that it is raining in Eketahuna as a consequence of looking at the tea-leaves in the bottom of my cup after breakfast in Dunedin, but I would not then *know* that it is raining in Eketahuna. True beliefs can be acquired as a matter of chance or luck, but knowledge cannot. This is not strictly true. Suppose I glance out the window and see Moana walking past. As a result, I come to believe that Moana is now in Dunedin. If that was the only time I glanced out the window all morning, and it coincided with the only time Moana passed by, then my acquiring that knowledge was a very chancy matter. But although I may have been lucky to acquire a true belief in those circumstances, it was not a matter of luck that the belief I did acquire was true. We must distinguish between cases where luck plays a role as to whether or not the belief is acquired and those cases where luck plays a role as to whether or not the belief is true. I shall refer to the latter as cases where the belief is *fortuitously* true.

According to the traditional analysis of knowledge, a further condition for knowledge is justification. More precisely:

> (TK)   S knows (at t) that *p* iff
> >   (a)   p is true
> >   (b)   S believes (at t) that p
> >   (c)   S is justified (at t) in believing that p.

Edmund Gettier (1963) has argued that these three conditions, although necessary, are not sufficient for knowledge. He offers examples in which, although (a), (b) and (c) are met, S fails to know that *p*. Adding the justification condition is not sufficient because justification does not rule out the possibility of a belief's being fortuitously true.

A typical Gettier counterexample is as follows. Smith is in a room with Jones and Brown. Smith knows Jones very well and has considerable evidence that Jones owns a Ford. (He has seen Jones driving a Ford, seen ownership papers, and so on.) Brown is a stranger to Smith. Smith has no evidence as to what sort of car, if any, Brown owns. Smith does not own a Ford. Smith forms the justified belief that Jones owns a Ford and from this deduces that someone in the room owns a Ford. But, in fact, Jones doesn't own a Ford. She sold it the previous evening. However, Brown does own a Ford. So the statement that someone in the room owns a Ford is true, Smith believes it and he is justified in believing it. But Smith does not know that someone in the room owns a Ford.[2]

Gettier points out that his counterexamples rely on the following two principles:

(G1)   It is possible for someone to have a justified false belief.

(G2)   If S is justified in believing that p, and S validly deduces q from p, and S accepts q as a result of this deduction, then S is justified in believing that q. (1963, p. 121)

What gives Gettier examples their 'bite' is (G1), the possibility of justified but false belief. To deny this possibility is highly implausible. There was a time when many people had evidence that justified their belief that the sun moves around a stationary earth. It is possible for a person to be justified in believing that a bank note in his possession is worth $20 when, in fact, it is counterfeit. In the Ford example, Smith is justified in believing that Jones owns a Ford even though Jones does not own one.

The Gettier examples describe situations in which the evidence for p is compatible with both p and $\sim p$. Now in such cases, if the evidence justifies believing p when it is true, why should exactly the same evidence not justify believing p when it is false? What could explain the difference? If the truth of p must enter into the explanation of why the belief is justified, it could only be because the evidence must entail p. If that were so, then our standard of justification would be so high as to require absolute certainty.

This demonstrates why it is not enough to reject the Gettier examples on the grounds that they only work because they rely on too low a standard for belief justification. No matter how high one's standard of justification, if that standard allows the possibility of justified false belief, then it will be possible, in principle, to devise a Gettier counterexample.

(G2) is not so plausible, but it is stronger than is required.. Perhaps not all valid deductions transmit justification. Long and complicated ones may not, but simple ones, like that carried out by Smith in the Ford example, surely do. All that is required is the possibility of justification being transmitted in such simple cases. A weaker version of (G2) would serve Gettier's purpose:

---

[2] Adapted from Lehrer 1965, pp.169-70.

(G2′)   If S is justified in believing that p, and S validly deduces q from
          p, and S accepts q as a result of this deduction, then it is
          possible that S is justified in believing that q.

It has been suggested that conclusions deduced from false premises may
not always be justified (Meyers & Stern 1973, pp. 148-52). Alternatively, it
has been suggested that 'general' conclusions (in the sense that they can be
true in many different ways) may not be justified if they have been deduced
from facts that are not, in fact, responsible for their truth (Thalberg 1969, pp.
798-99). But these suggestions would be better expressed as claims that we
may not obtain knowledge under such circumstances, even though our
conclusions are justified. And that, of course, would amount to agreeing with
Gettier. To deny (G2′) can apply in such cases is as implausible as denying
(G1) and would be analogous to doing so.

Besides, it is possible to devise Gettier examples that do not appear to
rely on (G2) or (G2′), nor on deduction from false premises. For example, I
know my aunt to be an honest and reliable person. She phones me from Eketahuna one morning and informs me that it is raining in Eketahuna. But in
fact, my aunt is confused (as a result of the pharmacist making a mistake
with the prescription for the pills that she took the previous evening). She
has no idea what the weather is like in Eketahuna this morning (not having
looked). Instead, she is remembering what the weather was like on the
previous morning. I justifiably believe that my aunt has always been honest
and reliable in her dealings with me and that she has just told me that it is
raining in Eketahuna. From those true premises I immediately infer that it is
raining in Eketahuna. In fact, it *is* raining in Eketahuna this morning, so I
have a justified true belief that it is raining, but I do not know that it is
raining. This example does not rely on (G2) or (G2′), because my reasoning
is more plausibly inductive rather than deductive. Nor does it involve inference from false premises. All of the premises used in my inference are true
(cf. Feldman 1974).

That example is unlikely to find favour with strict deductivists who
maintain that deduction is the only valid form of inference. But deductivists
must agree that there can be non-inferential knowledge, otherwise no sound
deductions that yield knowledge would be possible. Gettier counterexamples
that do not appear to involve any inference at all are also possible. I see what
appears to be a vase on the table. I have no reason to believe that I am being
deceived and have plenty of reason to believe that I am not being deceived. I
form the belief that there is a vase on the table. But what I am seeing is the
reflection (or perhaps hologram) of a vase that is not on the table. However,

there *is* a vase on the table that I do not see. So I have a justified true belief that there is a vase on the table but I do not know that there is a vase on the table (cf. Goldman 1967, p. 69).

It appears that the Gettier problem should be taken seriously. But there is an argument to the effect that although it should be taken seriously it should not be taken too seriously. The suggestion is that the search for a set of necessary and sufficient conditions for knowledge is misguided. The Gettier examples indicate that, rather than there being some essence of knowledge, instances of knowledge share at most a family resemblance (Saunders & Champawat 1964, p 9).

Without further explication, it is difficult to evaluate the worth of this argument. I can understand that there may be no essence of games which links rugby, chess, patience and hop-scotch, and similarly, no essence which is common to knowing that it is raining, knowing how to cook ratatouille, and knowing Angela D'Audney. But it is not so obvious that there are no conditions (perhaps vague conditions) that pick out rugby union, rugby league, soccer, gridiron, etc. as games of football. Perhaps we could even specify conditions for a game's being a competitive team sport. Similarly, it is not obvious that we should be unable to specify conditions for factual knowledge; or failing that for different types of factual knowledge, e.g. singular facts or general facts; or perhaps for knowledge obtained in different ways, e.g. by direct perception or from witnesses.

Another reason for not taking the Gettier examples too seriously might be to argue that they describe situations so bizarre that it is not surprising that our usual concepts cannot cope. A similar argument may be used to dismiss examples of person-splitting and person-joining when they are used to counter particular theories of personal identity. But the Gettier examples are not particularly bizarre and even more mundane ones can be devised. I take it to be an everyday event that we come to believe something, and justifiably so, on the authority of another. For that authority to be mistaken or to intend to mislead is always possible. And for the authority to be correct by accident (right for the wrong reasons) is also always possible. Those are the ingredients for a Gettier counterexample. Any artificiality in the examples need only have been introduced to clarify the issues involved.

The Gettier problem should be taken seriously. What is the lesson to be learned from the Gettier counterexamples? They teach us that it is not just unjustified beliefs (such as beliefs based on reading tea-leaves) that can be fortuitously true. It is possible for a justified belief to be false. With this possibility comes another: the possibility that in circumstances similar to those which would result in a false belief, that belief could be fortuitously true.

That is what happens in the Gettier counterexamples. And this directs our attention to the important lesson to be drawn from those counterexamples. Any condition on our beliefs that does not guarantee their truth will not provide a condition for knowledge that is sufficient to accommodate the fact that a fortuitously true belief cannot be knowledge.

4.4. EPISTEMIC VALUE

The Gettier lesson does not just concern the inadequacy of the justification condition. Which notion, or notions, legitimately counts as epistemic justification is controversial, but that controversy need not concern us here. When a belief is described as being justified (for the believer) then some epistemic value is being ascribed to that belief. Either the believer has good grounds available for the belief, or the believer can give good grounds for the belief, or the believer has not violated any epistemic duties in acquiring it, or the believer has acquired it via a reliable method, or the belief coheres with the believer's other beliefs, and so on. The point is that an ascription of epistemic justification (whatever one's notion of justification) is an ascription of epistemic value. The notion of epistemic value is wider than that of justification. We may accept that some property is epistemically valuable even though we would deny that it provides justification. For example, many (but by no means all) epistemologists would agree that being caused by a reliable process is epistemically valuable but deny that this provides justification (unless the believer is aware that it has been so caused). There are epistemic values to which no one would be likely to attach the label of justification. Even if we restrict our attention to epistemic values that are truth-conducive, this may still be the case.

By a *truth*-conducive value, I mean any value that makes a belief more likely (in some sense) to be true. There are other properties that a belief may have which we may value. Certain kinds of belief may bring comfort or pleasure to the believer but such properties need not correlate highly with truth. But when considering values that are relevant to the notion of knowledge, it is only the truth-conducive values that are of interest. Therefore, I restrict my use of the term 'epistemic value' to truth-conducive properties that a belief may have.

Another restriction is that epistemic value applies to beliefs *qua* mental states, not beliefs *qua* belief-contents. Generally we think of beliefs as a certain kind of mental state. They belong to the class of propositional attitudes along with desires, fears, doubts, suppositions, etc. As such, they have propositional content, that is, the content of each belief-state is a pro-position (whatever that may amount to). Often we use the term 'belief' to refer simply to the content of a belief-state. When we say that two people have the same belief, we usually mean that they have beliefs with the same belief-content (i.e. they believe the same thing). As mental states, their beliefs may have widely differing properties. One may be rational, the other irrational; one certain, the other tentative; one obtained by inference, the other by authority, and so on. On the other hand, their contents will have identical properties, and these will be semantic properties. They will have the same truth-value, meaning, and reference.

Epistemic values are not semantic properties. It follows that for any epistemic value it is possible to have false beliefs with that value, no matter how high the degree of value. It is sometimes claimed that it is impossible for a self-evident belief to be false. But as I have already argued (see Sections 3.6 & 3.7), this claim arises from an equivocation on whether self-evidence is a psychological property (one that applies to a belief state) or a semantic property (one that applies to a belief-content). If it is a psychological property then its applying will not guarantee truth and if it is a semantic property then its applying will not guarantee any particular psychological attitude to the belief.

Which values are most worthy is a matter of dispute. The major dispute is between the internalists and the externalists. Internalists put great importance on the ability of the believer to have access to the epistemic value in question. We can assess whether or not our beliefs have good grounds (so long as what we mean by 'grounds' is restricted to those things to which we have cognitive access), or whether they cohere with our other beliefs, or whether they have been validly inferred, and so on. Externalists, on the other hand, do not restrict themselves to what the believer can be cognitively acquainted with. We cannot always be aware of how reliable the method employed to acquire a belief may be. Nor can we always be aware of what the proper function of our cognitive faculties may be and thus whether a belief was acquired by properly functioning cognitive faculties. And so on.

The Gettier problem applies irrespective of what one takes to be the important epistemic value or values. For 'epistemic value' you may plug in any epistemic value that you consider to be necessary for knowledge.

External epistemic values, such as reliability of the process of belief acquisition, are just as subject to the Gettier problem as internal values. A reliable process cannot guarantee truth, unless it is 100% reliable. It is unlikely that we have any processes available to us that are 100% reliable. If knowledge could only be acquired via such processes then we would have far less knowledge than we have according to our intuitions. Telephoning my aunt is a reliable way of acquiring true beliefs about the weather in Eketahuna, but not all true beliefs acquired by that method are items of knowledge. It might be objected that a process that involves my aunt listening to sounds on the window, rather than looking out of the window, is an unreliable process. But I don't see why such a process should not be highly reliable, and yet occasionally lead to a false belief and even less often a fortuitously true belief.

One way of identifying analyses of knowledge that can be 'gettierised' is that the truth condition must be included as a separate condition. This gives the set of beliefs that are both epistemically valuable and true a certain lack of cohesion. It is as though the truth condition is tacked on to ensure that no false beliefs can count as knowledge. The set of true beliefs forms a cohesive set as does the set of epistemically valuable beliefs, but there are members of their intersection that lack a connection between their truth and their truth-conducive property.

4.5. THE STRONG CAUSAL CONNECTION

A typical inclination when faced with the Gettier problem is to suggest that what is required to solve it is a fourth necessary condition for knowledge that, combined with the three conditions of the traditional analysis, will be jointly sufficient for knowledge.

What appears to be lacking in each of the Gettier examples, but which we would expect to be present in an example of genuine knowledge, is a connection between the evidence that justifies S's belief that p and the fact that p. In the Ford example, there is no relevant connection between the fact that Brown owns a Ford and the observations on which Smith bases his conclusion that someone owns a Ford. Similarly, in the Eketahuna example, there is no connection between my aunt's report (which I believe) and the fact that it is raining.

So it is initially plausible to suppose that what is missing from the traditional analysis of knowledge is a condition that there be an appropriate connection between the fact and the evidence, and hence, between the fact and the belief. If it is the evidence which gives rise to S's belief that p, it had better be the fact (or state of affairs) that p which gives rise to that evidence. Further, it is reasonable to suppose that the sort of connection we are talking about is a causal connection. After all, it is causal connections that make

things happen 'out in the world'. This suggests a fourth condition for knowledge:

(C1)   The fact that p must cause S's belief that p (Goldman 1967).

If we consider the acquisition of knowledge by perceptual processes as
paradigmatic of knowledge acquisition, then such a condition seems a
natural one.

In the next chapter I discuss possible counterexamples to the claim that a
causal condition is a necessary condition for knowledge. But as formulated it
is clearly too strong. An immediate difficulty is that it denies knowledge of
facts about the future. There may be circumstances in which I can know that
I shall play billiards tomorrow.[3] But, if so, the fact that I shall play billiards
tomorrow cannot be the cause of my present belief that I shall play. Even if
backward causation is possible, it is clearly not operating in such cases.

This problem does not just apply to future facts. For example, when I am
sitting in front of the fire in my sitting room I may know that smoke is rising
from the chimney, but the fact that the smoke is rising is not a cause of my
belief that the smoke is rising.

Recall that our initial intuition was that an appropriate *connection* was
missing in the Gettier examples. So (C1) can be modified to simply require a
causal connection between fact and belief, by which we shall mean either
that the fact is a cause of the belief or that the fact and belief have a *common
cause* (Goldman 1967, p. 364). In the billiards example, the common cause
is my intention to play billiards tomorrow. My intention is both the cause of
my playing billiards tomorrow and the cause of my present belief that I will
play billiards tomorrow. Similarly, the fact that the smoke is rising and my
belief also have a common cause, namely the fire.

So we have:

(C2)   The fact that p is causally connected to S's belief that p.

Note that adding (C2) to the traditional analysis (TK) allows us to deal with
Gettier examples in the case of future facts. Suppose I intend to play billiards
tomorrow and justifiably believe that I shall do so. But tomorrow morning I

---

[3] I say 'may be' in deference to those of a more sceptical bent who would deny that there
could be such circumstances. But to deny the possibility of any knowledge of the future seems
over-sceptical. If I can know that I am now at the University of Otago, then surely I can know
that I will be in Dunedin in one minute's time. Anyway, a theory of knowledge should not
rule out by fiat the possibility of knowledge of the future.

learn that my usual billiards opponent has been hospitalised so I change my mind and set out to visit him instead. But on the way I am kidnapped and taken to the home of an eccentric *mafioso* who forces me at gunpoint to play billiards. The conditions of the traditional analysis are met, but I do not have knowledge because there is no relevant common cause of my belief that I shall play and the fact that I do so.

The claim that (C2) is a necessary condition for knowledge is, of course:

(SC)   S knows that *p* only if the fact that *p* is causally connected to
        S's belief that *p*.

This requirement that there be a causal connection between fact and belief promises to rule out the possibility of a belief's being fortuitously true. There are other advantages of such a causal requirement for knowledge.

4.6. AVOIDING FOUNDATIONAL REGRESS

A regress threatens if one adopts the notion that a person's beliefs are justified by other beliefs. For example, my belief that my cousin is in Kiribati may be justified by my belief that I have received a letter from her postmarked 'Kiribati'. This belief, in turn, stands in need of justification and this gives rise to an infinite regress. The foundationalist response is to claim that this regress can only be stopped if there are basic or foundational beliefs that are self-justifying—the holding of such a belief being sufficient for that belief to be justified.

Traditionally, foundational beliefs have been those which are certain or beyond doubt. Candidates for this role have been beliefs of pure reason (logical and mathematical truths) and beliefs about one's current sensory experiences and mental states. Traditional foundationalism faces two problems. First, it can be argued that the beliefs proposed as foundations are in fact subject to doubt and therefore are not self-justifying. Secondly, even if they are free from doubt, such 'etiolated' beliefs as Descartes' *Cogito* or 'I seem to be experiencing green, now' cannot justify more substantial beliefs about the external world.

The coherentist alternative, according to which each belief in a coherent set of beliefs is justified by the other beliefs in that set, simply replaces a vicious regress with a vicious circle or, more accurately, with a network of vicious circles.

With the causal-connectionist approach, we can eschew traditional justification by claiming that we have knowledge when the right sort of causal connection exists between belief and reality. Beliefs may be inferred from other beliefs but the regress stops with beliefs with an appropriate causal history. Whether or not we call such basic or non-inferential beliefs 'justified' is a terminological matter, but it would be inappropriate to call them 'self-justified'.

Insofar as the regress of justification gives rise to scepticism, the causalist doctrine suggests a solution to that problem. We have knowledge when our beliefs are appropriately connected to the facts, and this connection may be in place irrespective of whether or not we believe or are justified in believing that such a connection is in place.

4.7. PARADIGMS OF KNOWLEDGE ACQUISITION

The paradigmatic processes of knowledge acquisition such as sense perception, memory, and testimony are causal processes. Furthermore, it is fundamental to straightforward examples of such processes that knowledge acquisition is achieved when a fact gives rise to a belief in that fact, via the process. We cannot perceive, remember, or be informed of a fact, so as to know it, unless that fact plays a causal role in our perception, memory or information acquisition. That is what distinguishes hallucinations, pseudo-memories, and misleading testimony from genuine perceivings, rememberings, and reliable informings.

I say more about these paradigm processes, particularly sense perception, in the next chapter.

4.8. THE CAUSAL CONNECTION IS NATURALISTIC

(SC) is a naturalistic condition on knowledge. It does not involve normative epistemological notions of justification, rationality, and the like. For anyone who believes that a naturalised epistemology is a 'good thing', it is a promising starting point for such an epistemology. On the other hand, further conditions could be added to (SC), if deemed necessary, to yield a non-naturalistic account of knowledge.

I believe that (SC) forms the basis of a plausible account of knowledge. By itself it is not quite enough. An account of knowledge based on (SC) alone would be:

(M)   S knows that p iff S has a belief that p which is causally
           connected to the fact that p.

This minimalist causal theory of knowledge results from replacing the justification condition of the traditional analysis with the causal condition. Note
that a separate truth condition is now redundant. This gives the account a
cohesion that the traditional account lacked. Truth is no longer 'tacked on'
by stipulation, but fits into place as a necessary part of the causal connection.

Epistemic justification is a troublesome notion. To begin with, the term
appears to cover a raft of different notions. It is controversial whether only
one of those notions is the only true notion of justification or whether 'justification' is a generic term that legitimately covers a number of different notions or whether it is an incoherent notion that appeals to conflicting desiderata (Alston 1989 ch. 5 and Plantinga 1990). To describe a belief as being
justified (for the believer) might mean that the believer has good grounds
available for the belief, or the believer can give good grounds for the belief,
or the believer has not violated any epistemic duties in acquiring it, or the
believer has acquired it via a reliable method, or the belief coheres with the
believer's other beliefs, and so on. It is the internalist notions that are
traditionally associated with the term, especially deontological notions involving epistemic duty. And it is just those notions that would appear to be
anathema to a naturalised account. A further problem for the deontological
notion is that it conflicts with belief involuntarism. If we cannot choose our
beliefs, then it is difficult to see how we could be charged with failing in our
epistemic duties when acquiring beliefs.

(M) is tempting because it eschews justification. It may even be that (M)
comes close to satisfying our intuitions as to what counts as an item of
knowledge. The same cannot be said for all those items that meet only some
of the other (supposedly) necessary conditions for knowledge. Many justified beliefs are definitely not knowledge because they are false. Many true
beliefs are not knowledge because it is a matter of sheer chance that they are
true. Similarly for justified true beliefs as the Gettier examples show. To
show how remote from knowledge a justified true belief could be, suppose
our beliefs are caused by an evil but occasionally careless demon. Amongst
all our justified false beliefs would be the occasional justified true belief,
inadvertently instilled. Those beliefs would be far from counting as items of
knowledge.

The following scenario demonstrates that it may be appropriate to ascribe
knowledge in the absence of any justification so long as there is a causal
link. Suppose an omniscient God were to instil true beliefs in the mind of

Nicky, while she slept. Suppose further that she wakes convinced of the truth
of those beliefs but with no idea of their origin, and continues to believe
them without further evidence. God is the common cause of Nicky's beliefs
and of the world that those beliefs are about. Should we say that Nicky has
acquired knowledge? It is not clear. Our intuitions pull in different directions. Such beliefs would be totally unjustified from Nicky's perspective.
This inclines us to say that she does not really know, because she has no idea
of what is going on. But on the other hand, what better source of knowledge
could there be than an omniscient God? This inclines us to regard such beliefs as quintessential knowledge, providing a standard against which all
other knowledge acquisition could be judged.[4]

4.9. THE INSUFFICIENCY OF A MINIMALIST CAUSAL THEORY

But all beliefs that meet the conditions of (M) are not necessarily items of
knowledge or even close to being items of knowledge. Suppose that at other
times when she is asleep, Nicky is the victim of Satan who instils many false
beliefs. On waking she is just as convinced of those beliefs as those that are
instilled by God. Are we still inclined to say that she has acquired knowledge from God, particularly if the false beliefs far outnumber the true? We
might still insist that the God-instilled beliefs are cases of knowledge. But
even if the causal condition is not sufficient for knowledge, it does not
follow that the account needs to be augmented by the addition of non-
naturalistic conditions. In order to indicate that the causal condition can be
augmented by naturalistic conditions to yield a naturalised epistemology, I
shall outline some typical counterexamples to the sufficiency of the causal
account. There are familiar examples of knowledge being denied by the
existence of relevant alternatives in the vicinity. For example (Goldman
1976, p. 778), Jim sees Judy and forms the true belief that he has seen Judy.
But it could just as easily have been her twin sister Trudy. If it had been, he
would still have believed it was Judy, so Jim does not know. There are also
examples where knowledge is denied by the fact that the causal connection
is bizarre or deviant in some way. For example (Plantinga 1988, pp. 30-31),
there is the brain tumour that has the effect of causing the sufferer to believe
(in the absence of any other symptoms) that she has a brain tumour. Or
suppose that my usually reliable aunt tells me that it is raining in Eketahuna
while in a confused state due to a misprescribing of the pills she has taken.
Her pharmacist made the mistake on the previous evening because he was

---

[4] 'What a great way to learn a foreign language!' – Charles Pigden, in conversation.

suffering from a migraine that was caused by the climatic conditions (low barometric pressure, high humidity) prevailing at that time. Suppose, in turn, that it was those climatic conditions that were the cause of the rain that was falling at the time I spoke to my aunt. Now we have a causal chain connecting, via a common cause, my (justified) belief that it is raining with the fact that it is raining. But it would be odd to claim that I know that it is raining.

How can we rule out such cases? The causal connection will have to be an 'appropriate' one, and an account will have to be given as to what constitutes an appropriate connection. One approach is to use the notion of proper function (Plantinga 1993). Roughly, the proviso is that a causal connection can only provide knowledge if it is part of a process one of whose proper functions is to provide true belief, and that process must be functioning properly and reliably in an appropriate environment. A process can acquire a proper function either by intentional design or by natural selection. It is not clear that this approach could effectively deal with the Judy-Trudy case. The causal process of visual perception employed seems to satisfy the proper-function requirement, yet it is not a case of knowledge. One might argue that an environment that contains identical twins is not appropriate for the proper functioning of vision, or add the proviso, as Plantinga does, that there should be no 'glitch' in the environment (1993, p. 35).

Another approach is to suggest that a causal connection can only provide knowledge if it is part of a reliable process. This faces the problem, among others, of individuating reliable processes. If the process in the Judy-Trudy case is visual perception in good light, then it seems to satisfy the reliable-process requirement. One might suggest, as Goldman (1976) does, that it is the presence of a 'relevant alternative' in the vicinity that renders the process unreliable in this case.

Perhaps an amalgam of the two approaches will turn the trick, or perhaps even further additions are required.[5] My task is not to provide an account of the sufficient conditions for knowledge. Indeed, such a task is likely to prove a vain one. Over and above certain necessary conditions, different contexts call for different notions of what suffices for knowledge. What I am suggesting is that a causal-connection requirement provides the basis for a plausible naturalised account of knowledge.

---

[5] It is not clear that the two approaches are clearly distinguishable (see Taylor 1991). For the remainder of this chapter, I use 'reliabilism' as a coverall term for such approaches.

One response to this suggestion is to argue that an appeal to reliability threatens to turn my account of knowledge into a reliability theory — one in which causal-connection drops out of the picture. The reasoning behind this is that the causal connection was introduced to solve the Gettier problem, but fails to do so (as examples like that of the Ekatahuna pharmacist show). However, a reliable-process condition not only solves those for which an unadorned causal connection is insufficient, but promises to solve all the others as well. Thus, instead of adding the causal condition (C2) to the (justified) true belief conditions, we should add the reliable-process condition instead. But if we do that, then we can no longer claim that a causal connection is necessary for knowledge (cf. Wright 1983, p. 94 and Hale 1987, pp. 86-88).

But it is a mistake to think that reliability can shoulder all the responsibility for countering Gettier examples. Causal-connection may recede into the background, but it does not drop completely out of the picture. We can devise Gettier examples in which a reliable process gives rise to a true belief, but the belief is not knowledge because of the lack of a causal connection. Consider a thermometer with an exceedingly high (but not absolute) degree of reliability. In order to maintain this high degree of reliability the thermometer has been designed with a self-checking mechanism that operates for 10 seconds every 300 hours, on average. During those 10 seconds the thermometer shows a random number on its digital display. So the reliability of the thermometer is greater than 0.99999. Whenever Brenda, whose cognitive faculties are in reliable working order, looks at this thermometer (always under optimal conditions) she acquires the belief that the temperature is that which is displayed by the thermometer. Apart from those rare occasions when the self-checking mechanism is operating, she will acquire a true belief caused by a reliable process. When the self-checking mechanism is operating she will acquire false belief caused by a reliable process. Except, that is, for those extremely rare occasions when the random number displayed is indeed the correct temperature. Then she will acquire a true belief caused by a reliable process. But she will not *know* what the correct temperature is. Her belief is fortuitously true because it lacks a connection with the actual temperature.

Such examples confirm the 'Gettier lesson'. For knowledge we need a condition on beliefs that guarantees truth. Only a connection between reality and belief can achieve this, and an appropriate causal connection is the most plausible.[6] Revising the reliability condition so that it rules out the thermometer example without explicit mention of causal connection is not enough to throw doubt on the necessity of a causal connection. In the cases when Brenda does know the temperature there is still always a causal connection. So long as we lack examples in which a reliable process closes the gap between (justified) true belief and knowledge in the absence of a causal connection, we lack grounds for rejecting the causal condition in favour of a reliability condition.

In the previous section, I suggested that the causal-connection requirement offers the promise of a naturalised account of knowledge. But if the minimalist account (M) is insufficient without some sort of appropriateness condition on the connection, then it may seem that this promise cannot be fulfilled since 'appropriateness', of whatever variety, will be a normative notion. This concern is mistaken. Not all normativity is irreducibly non-naturalistic. Naturalistic explications of degrees of reliability, proper functioning, even rationality have been proffered. I shall not describe or assess them here. It suffices for me to point out that combining a strong causal condition with a suitable and successful explication of appropriateness presents the possibility of naturalising epistemology.

It is the deontological notion of justification that appears most resistant to naturalisation, but even that may succumb to some sort of 'means/end' analysis. Besides, a causal-connection requirement should still appeal to those for whom a naturalised epistemology has no appeal or for whom it is anathema. It provides a plausible means of avoiding the Gettier problem, a problem to which accounts relying solely on internalist notions of justification are so prone.

4.10. THE VIRTUES OF THE CAUSAL CONDITION

To sum up. Among the virtues of postulating that a causal connection is a necessary condition for knowledge are that it complies with our pre-theoretic notion of knowledge, it offers a solution to the Gettier problem, it is a naturalistic condition, it includes truth as a condition for knowledge without the need for stipulation, it avoids foundational regress, it is immune to certain varieties of scepticism, it does not depend on the truth of belief

---

[6] I examine, and reject, other 'connectionist' candidates in Chapter 6. Besides, the most plausible of those candidates do not allow platonic knowledge.

voluntarism, and it sits easily with paradigmatic examples of knowledge acquisition such as perception and memory.

CHAPTER 5

IS A CAUSAL CONNECTION NECESSARY FOR KNOWLEDGE?

5.1. OBJECTIONS TO THE CAUSAL CONDITION

(SC)   S knows that *p* only if the fact that *p* is causally connected to S's belief that *p*.

There are a number of objections to (SC). In this chapter, I examine each in turn. Some will prove more troublesome than will others. There are (at least) two things that I can hope to achieve in this chapter, one of which subsumes the other. Platonists have claimed that there can be no version of a causal constraint on knowledge that allows all non-controversial cases of knowledge, but disallows platonic knowledge (Steiner 1975, p. 116; Hale 1987, p. 100; Brown 1989, p. 110). Call a constraint on knowledge that allows all cases of knowledge, except platonic knowledge, a *viable* constraint. My more modest aim is to demonstrate that a viable version of (SC) is possible. In order to do so, it will be necessary to modify (SC) so as to counter some of the objections. The question that then arises is whether the modified version is as plausible as the original. It may be thought implausible because it is 'baroque' or unmotivated or *ad hoc*. My more ambitious aim is to justify each modification so that my final version of the causal constraint is plausible as well as viable.[1]

---

[1] Others may not use the terms 'viable' and 'plausible' in the same sense as I do.

5.2. PERCEPTUAL KNOWLEDGE

Perception is the fundamental mechanism by which we acquire knowledge of the world around us, and perception is a causal process. As Alvin Goldman puts it, drawing on the work of Grice (1961):

> [A] necessary condition of S's seeing that there is a vase in front of him is that there be a certain kind of causal connection between the presence of the vase and S's believing that a vase is present (Goldman 1967, p. 358).

That the causal connection is necessary is supported by the consideration of cases in which it is missing. Suppose there is a vase in front of S. But there is also a mirror between S and that vase and the mirror reflects the image of another vase, which is not in front of S. It appears to S that there is a vase in front of him. Although S may form the true belief that there is a vase in front of him, he does not *see* that there is a vase in front of him. In this case S lacks perceptual knowledge.

It might seem that I have assumed a realist account of perception, according to which 'S perceives X' entails 'X exists'. Some would dispute perceptual realism, claiming that in a case of hallucination, for example, someone may see a dagger before him although no such dagger exists. In my view, the hallucinator does not see a dagger. Rather, she has an experience *as if* she were seeing a dagger (see Musgrave 1993, pp. 276-77). However, to make my point I need not deny that the hallucinator sees a dagger in front of her. I need only claim that she does not see *that* there is a dagger in front of her. A causal, and hence realist, account of 'perception-that' need not conflict with non-realist accounts of 'perception-of'. Even the realist account of perception-that could be dispensed with. One could insist that the hallucinator sees that there is a dagger in front of her, but it would still be wrong to claim that she perceptually knows that there is a dagger in front of her. Similarly for S and the vases. Even if we concede (implausibly) that S sees that there is a vase in front of him, without the causal link, he still lacks perceptual knowledge.

Jaegwon Kim seems to deny that perception requires causation when he states that:

> [t]he condition that a causal relation obtain between the relevant items is just too broad a requirement to do any real work. ... [W]hatever further conditions are imposed on the simple causal condition will in fact have to do most of the work, and ... the causal condition itself will pretty much drop out of the picture (1977, p. 613).

But Kim's argument is essentially an argument for the insufficiency of causality in an analysis of perception (and hence, of perceptual knowledge). Certainly, not any causal connection will do. If someone tells me that there is a vase on the table behind me, then, although I may be causally connected to the vase via my informant, I have not perceived the vase nor do I have perceptual knowledge of it. The causal connection must be 'of an appropriate kind' for there to be a case of perception. We can agree that an account of this 'appropriateness' will do most of the work in a theory of perception, but this does not diminish the fact that the connection is, and must be, a causal connection. Kim's argument establishes only that the causal condition will fade into the background, not that it will drop out of the picture (Dretske 1977a, p. 623).

I assume that perceptual knowledge is non-inferential. Some epistemologists have maintained that perceptual beliefs are inferred from sense data or from sensory stimulations (e.g. Price 1961, pp. 394-97; Harman 1973, pp. 184-86). Such accounts need not violate (SC). If a certain fact 'appropriately' causes a sensory stimulation and a belief in that fact is inferred from that sensory stimulation, then, since inference is a causal process (as I argue in Section 5.3), the fact is causally connected to the belief in that fact.

Knowledge based on memory meets the causal condition since remembering is also a causal process. S remembers that p at time t only if S's believing that p at an earlier time is a cause of her believing that p at t (Goldman 1967, p. 360). Once again, the causal connection will have to be 'appropriate' if it is to be a genuine case of remembering.

5.3. INFERENTIAL KNOWLEDGE

Many of our beliefs arise by inference from other beliefs. For example, suppose I play a game of billiards with a stranger. Observing his play, I form the belief that he is a skilled player. From this, and certain background beliefs, I infer that he has spent a considerable time practising the game. So the fact that he has practised caused the fact that he is a skilled player, which causes my belief that he is a skilled player from which I infer that he has practised. In order to claim that there is a causal connection between the fact that he has practised and my belief that he has, we need to make the not unreasonable assumption that inference is a causal process whereby beliefs give rise to further beliefs (Goldman 1967, p. 362).

The following notation will be useful. [p] for the fact that p, B(p) for the belief that p, a solid arrow for a (non-inferential) causal connection, and a dashed arrow for an inferential connection. Let p be 'he has practised' and q be 'he is a skilled player', then Figure 1 illustrates the above example:

$$[p] \longrightarrow [q] \longrightarrow B(q) \dashrightarrow B(p)$$

Figure 1

Since inference is a causal process, my belief that p meets the causal-connection condition (SC).

5.4. GENERAL KNOWLEDGE

A difficult problem facing (SC) is that of coping with knowledge of a more general nature, in particular, knowledge of universal truths such as 'All crows are black' and existentially quantified truths such as 'Something is a black crow'. Universally and existentially quantified facts (at least those which are spatio-temporally unrestricted) do not seem suited to the role of cause or effect (Goldman 1967, p. 368; Hale 1987, p. 94).

Consider the fact that all crows are black. This fact may be cited in a causal *explanation* of a particular fact, such as the fact that a certain crow was shot, but it will not be cited as an actual *cause*. Rather it is particular facts, such as the fact that the crow in question was black (thus making it easier to be seen), that will be cited as causes. Universal facts may also be cited in explanations of other universal facts, as in 'Crows are easy targets in the snow because crows are black'. It might be argued that in such cases we have the 'because' of cause, not just of explanation. In other words, 'The fact that crows are black is a cause of the fact that crows are easy targets in the snow'. But agreeing that universal facts may be causes of other universal facts will not rescue (SC). (SC) requires that the fact that p be causally connected with the fact that S believes that p. The latter fact is a particular fact, and the concern is that universal facts do not cause particular facts.

A similar objection can be made with respect to existentially quantified facts. If I observe that Christine is grumbling and subsequently deduce that someone is grumbling, is there any good reason for asserting that it was the fact that someone was grumbling which caused my belief that someone was grumbling? Isn't it rather the fact that Christine was grumbling that was the cause (via an inferential link) of my belief, and whatever the causes of *that* fact, they will not include the fact that someone is grumbling?

Alvin Goldman suggests that we overcome this problem by adopting the following principle:

(L)   If p is logically related to q and $[q]$ is a cause of $[r]$, then $[p]$ is a cause of $[r]$ (1967, p. 368) where 'is logically related' means 'entails or is entailed by'. Although Goldman does not specify what he means by 'logically related', it is reasonable to assume this from his examples. (See Hale 1987, p. 95).

(L) accounts for knowledge of an existentially quantified truth as follows. I observe a particular bald man, say Lester. The fact that Lester is bald causes my belief that Lester is bald and from this I infer that there exists someone who is bald. So the fact that Lester is bald is a cause of my belief that someone is bald. Since the fact that someone is bald is entailed by the fact that Lester is bald, it counts, according to (L), as a cause of my belief that someone is bald.

Similarly, I can know the universal fact that all kiwis are flightless because that universal fact entails the content of each of my observations (or reliable reports) concerning flightless kiwis. At least Goldman seems to claim this (1967, p. 369), but it is far from clear that it is a case of entailment. I return to this point in Section 5.6.

Probably the most common criticism of Goldman's account concerns his idea that causal chains can be extended by logical links. The criticism has been not only that the principle is *ad hoc* and counterintuitive (Pappas & Swain 1978, p. 24), but also that it is false (Klein 1976, pp. 796-97; Hale 1987, p. 95).

The falsity of (L) is readily demonstrated. Suppose $[q]$ is a cause of $[r]$. Let p be any true proposition. Now p & q entails q, hence by (L), $[p\&q]$ is a cause of $[r]$. But p is entailed by p & q, so by (L), $[p]$ is a cause of $[r]$. (L) yields the result that every fact is a cause of any other fact.

(L) must be modified if it is to be retained. Recall that the motivation for (SC) is to identify the appropriate connection between fact and belief that is necessary for knowledge. A first step is to modify (L) so as to specify an 'appropriate connection' rather than a revised notion of cause:

(L′)   If p is logically related to q and $[q]$ is a cause of $[r]$, then $[p]$ is 'appropriately connected' to $[r]$.

It may be objected that this 'appropriate connection' is not really a *causal* connection, although in response it may be pointed out that such a connection does have an important causal element. A more serious objection is that to require such a connection between fact and belief would be a vacuous condition. By a similar argument to that above, (L′) yields the result that every fact is 'appropriately connected' to every other fact.

In order to accommodate general knowledge, the notion of causal connection needs to be weakened but not to that extent. The vacuity of (L′)

arises because it allows causal connections to be extended by entailment in both directions. I propose that only one sort of entailment relation is plausible and necessary. It is readily apparent that we can extend our knowledge by deduction. If S knows that p and p entails q and S correctly deduces q from p and as a result of this deduction S comes to believe that q, then this will be sufficient for S to know that q. Any theory of knowledge that denies the possibility of acquiring new knowledge in this way is too restrictive. The notion of an appropriate causal connection for knowledge should be extended so that S can know that p if S's belief that p is causally connected to a fact that entails that p. Thus:

(L'')   If p is entailed by q and $[q]$ is a cause of $[r]$, then $[p]$ is 'appropriately connected' to $[r]$.

It may be objected that one cannot *extend* one's knowledge by deduction on the grounds that knowledge is closed under deduction. In other words, if I know that p and p logically implies q then I know that q.[2] This is preposterous. There are logical implications of any proposition that are so long and complex that they could not be comprehended within a human lifetime. I cannot know something that I do not understand. But even if we accept this 'preposterous' view, it will in no way be in conflict with a version of (SC) which incorporates (L'').

On the other hand, a connection which involves entailment in the other direction will not always be a connection between a fact and a belief in that fact and therefore will not provide a condition of the sort required. S's belief that p could be causally connected to a fact that is entailed by p and yet p be false. A further stipulation that p be true would be required. I have already argued that a knowledge condition that does not 'tack on' truth is to be preferred (Section 4.4).

5.5. K-CAUSAL CONNECTION

It may seem inappropriate to continue to call such extended connections 'causal' connections. This problem is readily solved by devising a new term for what is a new concept. I shall adopt the term *k-causal connection* to denote the type of connection that is necessary for knowledge acquisition. A k-causal connection will connect a fact with a belief in that fact.

---

[2] '[N]othing ever was, or can be, proved by syllogism, which was not known, or assumed to be known, before.' (Mill 1843/1973, p. 183).

A preliminary definition of k-causal connection follows. In the accompanying diagrams, a double arrow denotes 'entails' and a single arrow denotes a causal connection (which may or may not include inferential links).

(K)   [p] is *k-causally connected* to B(p) iff either:

(a)   [p] is a cause of B(p), or

[p]   ⟶   B(p)

Figure 2

(b)   [p] and B(p) have a common cause, or

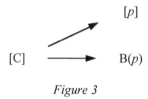

Figure 3

(c)   there is some *q* such that *p* is entailed by *q* and [q] is a cause of B(p), or

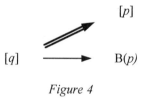

Figure 4

(d)   there is some *q* such that *p* is entailed by *q*, and [q] and B(p) have a common cause

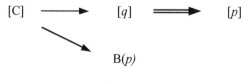

Figure 5

Condition (SC) now becomes:

(SC*)   S knows that *p* only if the fact that *p* is k-causally connected to
          S's belief that *p*.

(SC*) allows for knowledge of existentially quantified truths, but knowledge
of universal truths remains a problem. Before turning to that problem, the
notion of k-causal connection is in need of further explication.

My formulation of k-causal connection is an adaptation of Goldman's
notion that causal chains can be extended by logical-relatedness. I have restricted that relation to entailment in one direction. But what sort of entailment? After all, there is a cluster of notions variously labelled by terms such
as '(logical) entailment', '(logical) implication', '(logical) consequence',
'deducibility', and the like.

Suppose we take the logical relation for k-causal connection to be what
may be called 'strict' entailment. *p* is *strictly entailed* by *q* iff it is logically
impossible for *q* to be true and *p* false. This raises a problem with respect to
knowledge of truths that are necessarily true.

If k-causal connection involves strict entailment, then condition (SC*) is
trivially met in the case of necessary truths. This is because any necessary
truth is k-causally connected to a belief in that truth, so long as the belief has
a cause. Suppose S's belief in the necessary truth that *t* is caused by the fact
that *p*. No matter what the proposition *p* is, *t* will be strictly entailed by *p*. *t*
cannot be false, so *a fortiori*, it is impossible that *p* be true and *t* false. Figure
6 shows how the necessary fact that *t* is k-causally connected to the belief
that *t*.

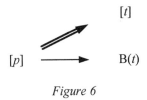

Figure 6

For example, suppose Lucy acquires the belief that either it is raining or it is
not raining. Suppose part of the cause of her acquiring this belief is a conversation with a logician. That she conversed with a logician entails that it is
raining or it is not raining. Thus her belief that it is raining or it is not raining
is k-causally connected to the fact that it is raining or it is not raining.

I am not suggesting that this sort of connection will be (anything like) sufficient for knowledge, simply that (SC*) would be met in the case of necessary truths because no matter what causes such a truth to be believed, there will always be such a connection. This looks like an advantage, since construing k-causal connection in this way does not rule out the possibility of knowledge of necessary truths.

Unfortunately, the weakness of this condition gives rise to a problem for the anti-platonist when we consider the possibility of non-tautological truths that are necessarily true. It is generally supposed that mathematics is necessarily true. If mathematical truths are necessarily true, then any true mathematical belief (which has a cause) will be k-causally connected to the mathematical fact that is believed. If that mathematical fact is about platonic objects, then (SC*) does not rule out the possibility of knowledge of such abstract objects. For example, suppose Susan believes that the number five is a prime number. If the platonistic construal of mathematics is correct, then her belief is true in virtue of the existence of a platonic object which exemplifies the property of primeness and if her belief was caused, say, by her mathematics teacher's pedagogical activities, then she is k-causally connected to a fact about a platonic object.

The anti-platonist response must be either to reject the claim that mathematics (at least when construed platonistically) is necessarily true, or to strengthen the notion of k-causal connection so as to rule out platonic knowledge, but without denying knowledge of logical truths (tautologies).

The claim that mathematics construed platonistically is, if true, necessarily true is questionable.[3] Alternative mathematical theories give rise to a problem for the claim. The Continuum Hypothesis (CH) is independent of the Zermelo-Fraenkel (ZF) axioms, but it and its negation are both consistent with the axioms (Cohen 1966). Is CH true of the realm of platonic sets that make the ZF axioms true, or is its negation? Some platonists, such as Gödel (1947/1983, p. 476), claim that it can only be one or the other. Suppose Gödel is right, and CH is the one that is true. Then one cannot help wondering if it might not have been true. Suppose the platonic sets had been somewhat different, so that ZF + ~CH were true of them. No inconsistencies follow from this (provided ZF is consistent), so why should it not be possible? But if this alternative realm is a genuine possibility, then platonistic mathematics is in some sense contingent.

Platonists must insist that whichever universe of sets exists, it exists necessarily. But if it is Cantorian set theory (ZF + CH) that is true in virtue

---

[3] The following argument is based on Cheyne 1989, pp. 16-17.

of the platonic realm, what is the status of non-Cantorian set theory? Is it genuine mathematics? Is there another realm for it to be true of? Two separate realms, or one realm that bifurcates at the transfinite? Most platonists do not tell us enough about this mysterious realm for us to know whether such questions even make sense.[4] They might claim that one theory is true of the platonic realm while the other theory is true in virtue of an if-thenist construal of it.[5] An obvious objection to this is that if one theory can be given an if-thenist construal, then why not all theories? An alternative approach could be to insist that set theory should be construed platonistically up to the bifurcation and given an if-thenist construal beyond that. This has some appeal, because one of the problems for if-thenism is that it seems to require a non-conditional knowledge base if it is to avoid complete arbitrariness.

This still would not erase the suspicion that there is something arbitrary about the claim that platonic sets have their existence and properties necessarily. We cannot help wondering, if such things exist, why they could not have been other than they actually are. Observe that most platonists adopt a quite different view when it comes to alternative geometries. They do not suppose that the abstract space in the platonic realm is either Euclidean or non-Euclidean. Neither do they suppose that there are alternative abstract spaces for the alternative geometries to be true of. Presumably this is because they believe that some sort of if-thenist construal is good enough for geometry. Why should it not be good enough for set theory? [6]

There are other arguments directed specifically against the claim that there are entities that exist necessarily. I explore these arguments in Chapter 7, which is concerned with existence claims as such.

If I am right and the substantive truths of platonism are contingent, then they are not strictly entailed by the sort of facts that cause beliefs. That being so, k-causal connection does not allow platonic knowledge. However, my argument against the necessity of platonic facts was directed specifically against mathematical platonism (assuming that all mathematics can be reduced to set theory). There are many other platonisms. Some of their non-logical truths may be necessarily true, although the sort of considerations I

---

[4] Plenitudinous or 'profligate' platonists grasp the nettle and assert that all possible sets exist in the one realm. I discuss plenitudinous platonism in Chapter 12.

[5] The if-thenist theory for construing mathematics is (roughly) the claim that all mathematical propositions are logically true conditional statements of the form 'A ⊃ T', where A represents the conjunction of the axioms of a particular theory and T any theorem which can be derived from those axioms. (See Russell 1919, p. 204.)

[6] Musgrave (1993, ch. 13) makes this point

have employed against mathematical platonism apply, at least *prima facie*, to other platonisms. But rather than strive for a global argument against the necessity of platonic truths, I turn to the plausibility of strengthening the notion of k-causal connection so that the k-causal condition is not met vacuously in the case of non-logical but necessary truths.

When I argued for the plausibility of extending causal connections with the entailment relation, I appealed to cases in which we extend our knowledge by deduction. This suggests that the relation required is that of derivability. Clause (c) of (K) would then be amended so that it requires that there is some q such that p is *derivable* from q and $[q]$ is a cause of B(p) with clause (d) similarly amended. But derivability is too strong. Which system of deduction is intended would need to be specified and any that is powerful enough would almost certainly be incomplete.

If strict entailment is too weak and derivability is too strong, then this suggests the intermediate notion of logical consequence. According to the standard (Tarskian) definition of logical consequence, sentence Q is a logical consequence of sentence P if and only if sentence Q is true in every interpretation in which sentence P is true (Tarski 1983). If 'P' stands for any sentence that expresses or denotes the fact that p (and similarly for 'Q' and q, etc.), then we can stipulate that the fact that q is a logical consequence of the fact that p iff there are sentences P and Q such that Q is a logical consequence of P. Logical facts will be logical consequences of any fact, in particular, of any fact that might be the cause of a belief. On the other hand, non-logical but necessary facts, if there be such, will not be logical consequences of such facts. There are many interpretations in which '5 is a prime number' is false, for example, when '5' denotes George W. Bush and 'prime number' applies to Hungarians. So if k-causal connection is explicated in terms of logical consequence, then logical truths will still meet the causal condition vacuously, while other necessary truths will be disallowed.

This appeal to the notion of logical consequence is not without problems for the anti-platonist. It appears to depend on the existence of sentences that may not exist. It may be a fact that p and S may believe that p but it does not follow that there exists a sentence P such that P expresses or denotes the fact that p. At least it does not follow that a token of P exists, and it will not serve my purposes to appeal to the existence of an abstract sentence type. But if beliefs are 'sentences in the head' and if S believes that p, then a sentence denoting p will exist. The brain state that instantiates S's belief that p will contain a representation of the fact that p, and that representation will be a sentence which denotes the fact that p. Whether that sentence is in a public language or in a language-of-thought ('mentalese') need not concern us (See Fodor 1976; Harman 1973, ch. 4; Lycan 1988, ch. 1).

Unfortunately, I wish to accommodate situations in which the fact that p is a logical consequence of the fact that q, which is in turn the cause of S's belief that p, as in Figure 7:

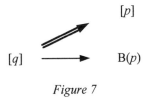

Figure 7

In such cases, there may be no belief that q nor any sentence Q. Appeal must be made to the *possibility* of a sentence existing. The definition of logical consequence must be altered so that q is a logical consequence of p just in case it is possible for there to be sentences P and Q such that Q is a logical consequence of P.

Another problem arises from the appeal to interpretations in my definition of logical consequence. In this context, interpretations are usually thought of as abstract mathematical entities. It makes no sense for me to appeal to the existence of such entities when putting forward an epistemic condition that purports to rule out the possibility of knowledge of such entities. But interpretations do exist as concrete tokens. So it is possible for interpretations to exist. I can redefine logical consequence in terms of such possibilities. Sentence P is a logical consequence of sentence Q iff it is not possible that there be an interpretation such that Q is true and P is false.

This reformulation of logical consequence now faces the objection that it appeals to possibilities. Are possibilities not platonic entities? Is not the notion of possibility only explicable in terms of platonic entities, whether they be the modal realist's possible worlds, maximal sets of consistent propositions, or whatever? But possibility, specifically logical possibility, has all the hallmarks of a *primitive* notion, in the same way that negation, conjunction, and existential quantification are primitive notions. Primitive notions are not understood by definition. Rather their meaning is conveyed by the procedural rules involved in inferring with them (Field 1989, pp. 32, 76). Certainly, we have a pre-theoretic notion of possibility that does not *prima facie* rely on the existence of platonic entities. Consider the following statements:

(a)   It is possible that there be a ten word English sentence
           containing the letter q at least five times.

(b)   There are no platonic entities such as numbers.

Someone asserting both (a) and (b) would not appear on the face of it to be
uttering a contradiction. Yet platonistic modalists would insist that (a) asserts
that some sort of platonic entity exists.[7] This shows that the platonistic
accounts do not capture our ordinary notion of possibility. Why give up our
ordinary notion without a compelling reason, such as a demonstration that it
involves a contradiction? Such a demonstration seems unlikely, given that
our ordinary notion is closely tied, if not identical, to the notion of absence
of contradiction.

David Lewis (1986) argues that rival theories to his modal realism all
appeal to some notion of primitive modality. But his account only avoids
such an appeal at great ontological cost. We are stuck with primitive modality. That being so, more sophisticated modal theories that speak of platonic
entities may be interpreted as speaking of possible platonic entities. I conclude that there is no convincing reason to suppose that the notion of logical
consequence is unavoidably committed to the existence of platonic entities.

A third problem with employing logical consequence along the lines that
I have suggested is that appears to be too strong. Consider the following
example. The fact that Arthur is a bachelor causes S to believe that Arthur is
a bachelor and from that belief S infers, and comes to know, that Arthur is
unmarried. But is S's belief that Arthur is unmarried k-causally connected to
the fact that Arthur is unmarried? That depends on whether or not the fact
that Arthur is unmarried is a logical consequence of the fact that Arthur is a
bachelor. Now, the sentence 'Arthur is unmarried' is not a logical consequence of the sentence 'Arthur is a bachelor'. But 'bachelor' is synonymous
with 'unmarried male', so the sentence 'Arthur is unmarried and Arthur is a
male' denotes or expresses the fact that Arthur is a bachelor. Hence, the fact
that Arthur is unmarried is a logical consequence of the fact that Arthur is a
bachelor.

My definition (K) of *k-causal connection* can now be expressed in terms
of logical consequence. But there is one further possibility that should be
allowed for. Suppose S's beliefs that q and r are k-causally connected to the
facts that q and r, respectively. Suppose, further, that p is a logical consequence of q and r, and S infers that p from q and r. It is not clear that S's
belief that p would meet any of conditions (K) (a)–(d), especially if one of

---

[7] Based on an argument of Field 1989, pp. 33-34.

the k-causal connections involves a common cause. If connections of types (a)–(d) are acceptable as the sort of connections between fact and belief that are necessary for knowledge, then combinations of such connections must also be acceptable. I propose that the notion of k-causal connection be extended to include such combinations of k-causal connections.

The relevant clauses of (K) may be revised so that we have:

(K*)   [p] is *k-causally connected* to B(p) iff either

   (a) [p] is a cause of B(p), or

   (b) [p] and B(p) have a common cause, or

   (c) there is some [q] such that [p] is a logical consequence of [q] and [q] is a cause of B(p), or

   (d) there is some [q] such that [p] is a logical consequence of [q], and [q] and B(p) have a common cause, or

   (e) there is some [q] and [r] such that [q] and [r] are k-causally connected to B(q) and B(r), respectively, and [p] is a logical consequence of [q&r].

and the revised causal condition retained as:

(SC*)   S knows that p only if the fact that p is k-causally connected to S's belief that p.

5.6. KNOWLEDGE OF UNIVERSAL FACTS

Extending the connection required for knowledge to k-causal connection, allows for the possibility of knowledge of existentially quantified truths. But can we account for knowledge of universal generalisations if k-causal connections are a necessary condition for knowledge? Recall that Goldman's suggestion was that I can know the universal fact that all kiwis are flightless because that universal fact entails each of my observations (or reliable reports) of flightless kiwis. See Figure 8, where p_i denotes 'kiwi i is flightless':

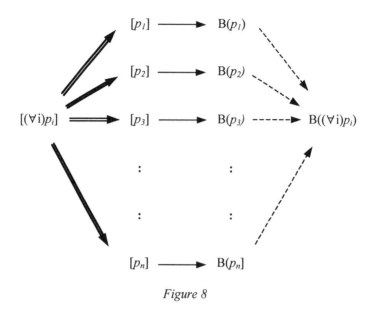

Figure 8

This shows a connection between the universal fact and my belief but it is not a k-causal connection because the entailment or relation of logical consequence is in the wrong direction. The universal fact entails rather than is entailed by the facts $[p_1]$ ... $[p_n]$.[8] Extending causal connections to include such a connection produces a vacuous condition as I have already shown.

But the notion of k-causal condition does not need to be extended in this way. I suggest that the existence of a common cause can account for our knowledge of universal generalisations. (This suggestion comes from Brown 1990, p. 114.) In the case of my knowing that all kiwis are flightless, each of the finite number of kiwis with which I have causal contact is causally connected back to some fact C (perhaps the fact that the 'first' kiwi was flightless) which is, in turn, a cause of the fact that each and every kiwi is flightless. See Figure 9:

---

[8] It is far from clear that it is a case of entailment or logical consequence. No matter, since I propose that the whole idea be discarded.

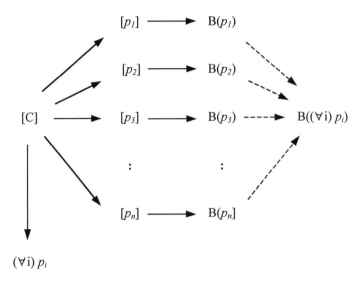

Figure 9

The notion of there being a common cause for the instances of a universal fact may seem peculiar to similar biological examples. In other cases the common cause need not be some fact involving the first instance. It is likely to be a complex fact involving a number of related aspects of the world. But whatever fact or complex of facts gives rise to the fact that every A is a B will also be a cause or partial cause of the fact that a particular A is a B. So, I can know that copper conducts electricity in virtue of my belief's being caused by those facts which cause it to be the case that copper conducts electricity, and so on.

Suppose we can make such an account plausible for cases such as the greenness of all emeralds, the flightlessness of all kiwis, the conductivity of all copper, and the like. Such accounts depend on there being some fact that is the cause of the universal fact. Unfortunately, it is reasonable to suppose that there are universal facts that do not have causes. Why should there not be universal facts that are fundamental 'brute' facts, true because that is the way the universe is and always has been? Consider the fact that all protons have a rest mass of 1.672614×10^{-24} grams. If protons have always existed and their rest mass is not further explicable, then this is an example of a universal fact that does not have a cause. Hence, it is a fact that cannot meet the k-causal-connection condition in the manner under consideration.

It is possible that there are no such facts. If there is a god that created the world and we can have appropriate causal contact with the singular facts about this god, which were causally responsible for the world being the way it is, then there will be no universal facts with which we cannot have causal contact. The same might be true if the world was caused by an uncaused big bang or if the 'Russian-doll' model of the world is true. (The 'Russian doll' model has it that every fact is caused by another fact and so on down ad infinitum.) These are, of course, contingent matters and a theory of knowledge ought not to rely on the truth of such scenarios.

One response to this objection is to point out that if the causal-connection condition denies knowledge of fundamental universal facts, then, although this yields scepticism about such facts, this is an acceptable scepticism that we can live with. So we cannot know the brute universal truths about the world? Why should we think that we ever could uncover and come to know such truths? Scepticism about brute universal facts need not be a problem, so long as it does not spread to other facts.

There may seem to be a problem for those of our inferences that depend on universal facts. If those facts are facts that cannot be known, can the conclusions we draw from such inferences be known? But we have little reason to think that any of our inferences depend on brute facts, certainly very few, if any, do. Why believe that scientific knowledge is anywhere near the fundamental level, if there is one? Scepticism about brute universal facts raises scepticism about the possibility of a completed science, but we have other reasons for being sceptical about that possibility anyway.

An alternative solution to the problem for k-causalists of knowledge of universal generalisations may be found by combining the theory that laws of nature are relations between properties (Armstrong 1983; Dretske 1977b; Tooley 1977) with a physicalist theory of universals (Armstrong 1978b). This solution also avoids the problem of knowledge of brute facts. Let P be the property of being a proton and let M be the property of having a rest mass of 1.672614×10^{-24} grams. If it is a law of nature that anything that is P is also M, then a certain relation of contingent necessitation N holds between the two properties. This yields the state of affairs N(P, M) (Armstrong 1983, p. 85). Furthermore, any proton a which instantiates the law yields the state of affairs (N(P, M)) (a's being P, a's being M). In other words, the law is a relation (dyadic universal) that holds between states of affairs (Armstrong 1983, pp. 87-88).

According to the physicalist theory of universals, universals are entities that cannot exist separately from the particulars which they instantiate. Furthermore, universals are physical entities that are spatio-temporally located. They are located at their instances. But what distinguishes universals from particulars, is that a universal is wholly present wherever it is instantiated. Universals are to be conceived of as repeatable features of the physical world. When a universal is instantiated, for example, when a rose is red, we have a physical state of affairs whose constituents are the particular rose and the property of redness. If this account is correct, then when we have causal contact with such a state of affairs, there is no reason to suppose that we do not have causal contact with the property of redness. After all, the fact that redness is instantiated rather than blueness has causal consequences (Armstrong 1978b; Bigelow 1988).

Now, if a law of nature is a (dyadic) universal and a universal is wholly present at all of its instances and we can have causal contact with the universal when we have causal contact with one of its instances, it follows that we can have causal contact with a law of nature without having causal contact with all instances of that law.

According to this theory of laws of nature, an example of a law of nature may be the fact that P-ness and M-ness are related in a certain way (although there may also be other kinds of laws of nature). A law of nature is not simply the fact that all Ps are Ms. The fact that all Ps are Ms might have been an accidental uniformity. However, the generalisation is a logical consequence of the law of nature, although it does not logically imply it. That is, $N(P, M) \Rightarrow (\forall x) (Px \supset Mx)$ but $(\forall x) (Px \supset Mx) \nRightarrow N(P, M)$. This means that although it may not be possible for us to be causally connected to the generalisation, we can be k-causally connected to it.

The account of laws of nature based on a physicalist theory of universals promises to provide the means for solving the problem of knowledge of universal facts that is faced by (SC*). But for many the price is unacceptably high. Universals conceived of as repeatable features of the physical world strike many as being 'metaphysically weird'. The idea that an entity could wholly exist at more than one place seems to involve a contradiction. If it does not, then it is not clear what being 'wholly present' means. Furthermore, nominalists (those who deny the existence of universals) suspect that the theory is poorly motivated since it arises from what they take to be a 'pseudo-problem', namely, the 'one-over-many problem (Devitt 1980). On the other hand, those more attracted to Plato's theory of universals will wonder how a particular could instantiate a universal if that universal did not already have independent existence. Suppose there was a time when there were no red objects. How could the first red object have instantiated redness if the property of redness was not already in existence waiting to be instantiated?

Although the Armstrongian theory of universals promises to provide a solution to the problem of knowledge of universal facts, the theory itself has many implausible and unsatisfactory features. On the other hand, the 'common-cause' solution is too sketchy to be entirely convincing. Knowledge of universal facts remains an obdurate problem for a causal-connection condition on knowledge.

5.7. THE BEHEADING COUNTEREXAMPLE

The following has been proposed as a counterexample to the claim that causal connections are necessary for knowledge (Skyrms 1967, pp. 385-86). I am walking along the street when I observe the decapitated body of the unfortunate Reid lying in the gutter. I have often seen Reid lying in the gutter before, as he is a notorious drunk. As a result of my observation, I come to know that Reid is dead. But unknown to me, Reid's death happened some time earlier as a result of a heart attack, which occurred when he was lying drunk and motionless in the gutter. Subsequently, a madman with an axe chanced upon the scene, saw Reid lying there, cut off his head, and departed before my arrival. My belief that Reid is dead is not causally connected with his death and yet I know that he is dead. It appears that a causal connection between fact and belief is not a necessary condition for knowledge.

But the causal connection the theory requires is between my belief that Reid is dead and the fact that Reid *is dead*, while the example speaks of a (missing) connection between my belief and his death i.e. the fact that Reid *died*.

Let t_H, t_B and t_O be the respective times of the heart attack, the beheading and my observation. Now, it is true that the fact that Reid is dead at t_H does not play a causal role in the acquisition of my beliefs at t_O concerning the ghastly situation. What I shall argue is that the fact that he is dead and my belief that he is dead have a common cause and are, therefore, causally connected (Loeb 1976, p. 331-34; Goldstick 1972, pp. 241-42). Although it is true that between t_H and t_B, the cause of his being dead was the heart attack, after t_B his being dead is causally over-determined. The causal over-determinants are both the heart attack and the beheading. Either fact alone would have caused Reid to be dead at t_O. The situation is analogous to Goldman's example of a fifth leg being placed beneath a table-top where 'it subsequently has a causal role in support of the table-top' (Goldman 1967, p. 362).

Some may be reluctant to accept that the causal overdeterminants of a fact are strictly causes of that fact. (Perhaps because neither is necessary, in the circumstances, for that fact.) But this is not sufficient reason for denying that each causal overdeterminant plays a causal role in the occurrence of that fact—one that is sufficient to provide the sort of causal connection that is necessary for knowledge. So 'cause' in (K*) should be understood to include both cause and causal overdeterminant.

Let p and q be 'Reid is dead at t_0' and 'Reid is beheaded' respectively. Figure 10 illustrates the situation:

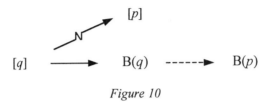

Figure 10

$[p]$ and B(p) have a common cause, *viz.* $[q]$, and are, therefore, causally k-connected. The beheading example is not a counterexample to (SC*).

It has been suggested that perhaps the example need only be modified to yield a genuine counterexample (Loeb 1976, pp. 334-35). Suppose, as seems inevitable, that as a result of my observation, I form the belief that Reid died. The fact that he died does not cause my belief, nor do that fact and my belief have a common cause—the fact was caused by the heart attack but the belief was caused by the beheading. So, it may be argued, although I can know that he died, there is no causal connection between the fact and my belief.

But this is to overlook the fact that k-causal connection includes causal chains extended by logical consequence. The fact that Reid died is a logical consequence of the fact that he is dead. By extending Figure 10 we see that there is a k-causal connection between the fact that Reid died and my belief that he died. Let p and q denote 'Reid is dead' and 'Reid is beheaded', as above, and let r denote 'Reid died'. Figure 11 illustrates the situation:

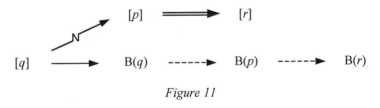

Figure 11

The modified beheading example is not a counterexample to (SC*).

There are problems with extending the notion of k-causal connection to include causal overdeterminants in this way. It may be objected that the beheading is not an overdeterminant, because it does not causally contribute to Reid's death. Only if the heart attack and the beheading occurred simultaneously would his death have been causally overdetermined. The beheading is rather a counterfactual cause of death and counterfactual causes are not actual causes. Extending k-causal connections to include counter-factual connections is problematic, not least because I criticise Nozick's proposal involving subjunctive conditionals in the next chapter (Section 6.2).

However, if k-causal connection with universal facts is possible (suppose one of the solutions discussed in Section 5.6 is on the right lines), a better solution is available. If I know that Reid died and is dead, it must be that I believe the nomic truths that all beheaded persons die and that all beheaded persons are dead. If my beliefs are k-causally connected to the respective facts and I am k-causally connected to the fact that Reid is beheaded, then my beliefs that he died and is dead are k-causally connected to the respective facts by clause (K*) (e).

The beheading example does not constitute a problem for (SC*) over and above the problem posed by universal truths.

5.8. NON-CAUSAL CORRELATES

Another problem or (SC*) arises from the claim that we can sometimes come to know that p by means of some other fact which, although it is a reliable indicator that p, does not bear any causal relation to the fact that p. For example:

> I recognise my uncle by his face although his having that face and his being my uncle are causally unrelated. His having that face, the one by means of which I recognise him, could be the result of a disfiguring accident. (Dretske & Enç 1984, p. 522)

To assess this objection, we must be clear about what it is that I am supposed to know. Is it:

(a)   the fact that the person I see is my uncle, or
(b)   the fact that I see my uncle, or
(c)   the fact that my uncle is in front of me,

or something else? In each case, it seems reasonable to claim that (SC*) is not violated. The fact that the person I see is my uncle causes my belief that the person I see is my uncle, the fact that I see my uncle causes my belief that I see my uncle, and the fact that my uncle is in front of me causes my belief that my uncle is in front of me. All these seem true of the situation described. Dretske and Enç object that specifying causes as facts obscures essential distinctions. They advocate adopting the gerundive mode in order to lay clear what they call the 'effective cause' (1984, pp. 518-19). Suppose my uncle's name is Jim. It is Jim's having a certain facial appearance which is the effective cause of my belief that I see my uncle, not Jim's being my uncle, anymore than it is Jim's being a carpenter or his having a hole in his left sock that is the cause. It is true that the fact that I see a man with a hole in his left sock causes my belief that I see my uncle if 'man with a hole in his left sock' is merely descriptive of Jim, the man with a certain facial appearance. But it is false if 'man with a hole in his left sock' is taken as identifying a property that plays an effective causal role in the acquisition of the belief. The same equivocation applies to the predicative expression 'my uncle'. In Section 4.2, I suggested that 'factive' causal claims are inter-translatable with 'gerundive' causal claims. Therefore, it must be the latter translation I have in mind and the fact that I see my uncle is not the cause of my belief that I see my uncle since it seems that it is not his being my uncle which is the cause of my belief.

But it is a mistake to suppose that Jim's being my uncle does not play a causal role in my successful identification. If I am to *know* that it is my uncle, then more will be involved than just my seeing someone with a certain facial appearance. I shall have to have some means of connecting that appearance with his unclehood. I can do this if I already have the belief that anyone who has that appearance is (or is highly likely to be) my uncle. In other words, I infer that it is my uncle from the belief that I see someone with a certain appearance plus the belief that anyone with that appearance is my uncle. It is this latter belief that is caused by the fact that that person is my uncle. To be more precise, the fact that the person was my uncle at some time in the past is a common cause of the fact that he is now my uncle and my belief that anyone with that appearance is my uncle which in turn causes (by inference) my belief that he is my uncle.

A possible objection is that my uncle-recognition is not a case of inference — that the process is too immediate for it to be that. Even so, I must possess some cognitive state that produces the belief that it is my uncle when I see him (perhaps I have acquired some sort of template in my mind, which automatically triggers uncle-recognition when I am appeared to in the requisite way), and that cognitive state must have been caused by the fact that he is my uncle. At least, it must if I am to have knowledge. If I don't know what my uncle looks like, I cannot know that it is my uncle when I see him, even if I believe it is for some bizarre reason.

The uncle-recognition example is not a counterexample to (SC*), but a similar example may be more problematic. Toby lives on an isolated island on which all yellow fruit contain citric acid. (There are no bananas on the island.) Citric acid curdles milk and curdled milk is a basic ingredient of the cheese that Toby makes. Toby believes that the juice of any yellow fruit will curdle milk and he has plenty of inductive evidence for this belief. One day Toby has some lemon juice in his hand and is about to add it to some milk. He believes that the milk will curdle. Indeed it seems reasonable to say that he *knows* that the milk will curdle (Dretske & Enç 1984, p. 526).

What is not clear is that the fact that the milk will curdle is causally connected to his belief that the milk will curdle. The cause of the fact that the milk will curdle is that the juice contains citric acid. The cause of Toby's belief is the fact that the juice came from a yellow fruit together with his general belief about yellow fruit.

As with the beheading example, a solution is available if k-causal connection with general facts is possible. If Toby believes that the juice of all yellow fruit on his island will curdle milk and his belief is k-causally connected with its corresponding fact, and also believes that the juice in his hand is the juice of an island-yellow fruit (and that belief is caused by its corresponding fact), then his belief that the milk will curdle is k-causally connected with the fact that the milk will curdle by (K*) (e). Once again, this example does not constitute a problem for (SC*) over and above the problem posed by universal truths.

5.9. KNOWLEDGE OF NON-LOCAL QUANTUM FACTS

Another proposed counterexample to the claim that causal connections are necessary for knowledge is based, not on thought-experiments, but on the results of actual experiments in physics (Brown 1990, pp. 114-17). The experiments are those that demonstrate the failure of the Bell inequality to be realised in an EPR-type setup (d'Espagnat 1979). When an energetic particle decays into a pair of photons moving in opposite directions, spin is conserved. If the spin along (say) the x-axis for one proton is measured as spin-up then the measurement along the same axis for the other proton will be spin-down. So by observing the spin measurement of one proton we can know the outcome of the other measurement without observing it or being causally influenced by it. Indeed, the possibility of causal influence can be ruled out by separating the measurements so that they are outside each other's light cones; in other words, so that no signal could pass from one measurement to the other in time to have an effect on it. On an EPR (or 'hidden variables') interpretation of quantum mechanics, knowing what the other measurement is without observing it does not violate (SC*). This is because on that interpretation the measurements are causally connected via a common cause. Which spins the protons have are caused in some way by the event that separates them at their source. I can come to know the spin measurement of the other proton in the same way that I can know by observing the fire in the grate that smoke is coming out the chimney or that I can come to know future facts.

On the alternative Copenhagen interpretation of quantum mechanics such an account will not do. According to that interpretation, the spin components are created by the measurements. There are no spin components or underlying variables that determine spin value until a measurement is actually made. (In other words, there is no common cause at the source.) But in order that spin be preserved, this must mean that the result of one measurement creates the result of the other measurement without causally influencing it.

Unfortunately for the EPR interpretation, it is the Copenhagen interpretation that is supported by empirical evidence. Assuming EPR and the locality assumption of special relativity, the Bell inequality can be deduced, and experiment confirms that this inequality does not hold. But this is far from providing a decisive (or even plausible) argument against the causal connection requirement for knowledge. All interpretations of quantum physics are controversial. Some respected physicists, including Bell, believe that special relativity needs to be modified. One alternative is the so-called 'holistic' view that the two protons should be regarded as a single system, although this may have the counter-intuitive consequence that this holism spreads to the universe as a whole. (Everything is interconnected.) James Brown suggests that a notion of non-physical cause could be invoked to save the causal theory of knowledge (1990, p. 117 fn.47). Others argue that by abandoning certain classical descriptions of reality (in particular, the notion that processes are continuous in space and time), we can preserve a causal account of the distant correlations (Chang & Cartwright 1993). I believe that it is our notion of physical cause that may need revision. In any event, interpreting the Bell results is too controversial to provide a compelling counterexample to the necessity of a causal connection for knowledge.

5.10. THE *TU QUOQUE* OBJECTION

Claims concerning conditions on knowledge may give rise to self-referential worries. It had better not be the case that a belief that the condition holds cannot, or is unlikely to, meet the terms of the condition itself. The claim will also be undermined if a belief in the condition gives rise to a contradictory belief.

I believe:

(SC*)   S knows that p only if the fact that p is k-causally connected to S's belief that p.

Accordingly, if I am to know that (SC*) is the case, then the fact that (SC*) must be k-causally connected to my belief that (SC*). Is this feasible?

The fact that (SC*), supposing that it is a fact, is a very complex fact about the world. On the face of it, it is a universal fact. Its formulation appears to quantify over propositions. It may be that it can be reformulated to quantify over only facts and beliefs. In other words, as the claim that all beliefs that are items of knowledge are beliefs that are k-causally connected to their corresponding fact. If the objection is that I do not or could not be k-causally connected to the fact that (SC*) because (SC*) is too complex or is a universal fact or has platonic entities as components, then the objection is not special to (SC*). These are problems that I have already attempted to deal with.

There is a further worry. (SC*) is a claim, not just about how things actually are, but about how they must be. This is not to suggest that (SC*) is a necessary truth. In possible-worlds talk, the claim is that (SC*) is true not just of the actual world, but of close possible worlds – worlds with epistemic agents and physical laws much like those in the actual world. So the fact that (SC*) is a modal fact. Once again, this is not a problem unique to (SC*). But is does make the problem of how we are k-causally connected to modal facts more acute. My purposes would be best served by an actualist account of modality that does not appeal to platonic entities or by an eliminativist account. There are no particularly appealing candidates, but no accounts of modality are free of serious metaphysical and epistemological difficulties (Sec. 5.5).

(SC*) concerns all beliefs that are items of knowledge. If the belief that (SC*) is to be an item of knowledge then that belief is self-referential. Some self-referential beliefs can only be held on pain of contradiction, or are false in virtue of their being held, or lead to some other absurdity. (This is not to say that holding such beliefs is impossible.) For example, the belief that all my beliefs are false, or the belief that I have no beliefs. Many self-referential beliefs do not have such consequences. For example, the belief that all my beliefs are known by God. So far as I can see, the belief that (SC*) is of the latter kind.

5.11. THE CONDITION OF THE CAUSAL CONDITION

My first aim for this chapter was to demonstrate that a modified version of the strong causal constraint on knowledge allows all non-controversial cases of knowledge while denying platonic knowledge. I claim that the k-causal condition:

(SC*)   S knows that p only if the fact that p is k-causally connected to S's belief that p.

achieves this. A number of purported counterexamples to strong causal conditions on knowledge have been advanced. (SC*) is viable insofar as it avoids such counterexamples. How plausible one considers (SC*) to be will depend on how plausible one considers physicalistic theories of universals and laws of nature, or the claim that universal facts and their instances have common causes, or the notion of causal overdetermination, and so on. I conclude that the causal constraint on knowledge emerges from its battle with counterexamples bloodied but unbowed.

CHAPTER 6

OTHER THEORIES OF KNOWLEDGE

6.1. CONNECTIONIST CONDITIONS ON KNOWLEDGE

In Section 4.4, I argued that any plausible theory of knowledge must include conditions that guarantee truth. Truth should not be tacked on as a further condition. This must be so in order to avoid the Gettier problem. I also argued that no internalist condition on knowledge could provide the necessary guarantee of truth. The only way that the truth of a belief can be assured is by a connection between fact and belief. In Chapters 4 and 5, I defended the claim that it is a causal link (more precisely, a k-causal connection) that is needed to make that connection. In this chapter, I examine alternative theories of knowledge that suggest a different sort of connection between fact and belief. In each case, I discuss whether or not the proposed connection is a necessary condition for knowledge and compare it with the k-causal condition.[1] I then discuss, putting objections to the particular theory to one side, whether or not the proposed connection allows knowledge of platonic objects.

6.2. THE TRUTH-TRACKING CONNECTION

The fact that p and S's belief that p may be connected according to the following subjunctive conditionals:

(T1)   If p were not true, then S would not believe that p

---

[1] This discussion draws on material in Cheyne (1997a).

(T2)   If p were true, then S would believe that p.

Robert Nozick (1981, pp. 172-78) claims that any belief that meets those conditions 'tracks the truth'. By this, he means that (on a possible-worlds analysis of subjunctive conditionals) in the possible worlds close to the actual world, both p and 'S believes that p' have the same truth-value.[2] Putting it another way, in all situations sufficiently similar to the actual situation, if p is true then S believes that p and if p is false then S does not believe that p. Since the actual world is itself a close possible world, given that S actually believes that p, it follows that p is actually true. Thus, the truth-tracking conditions pick out beliefs that are true.

Expressed as a necessary condition for knowledge, the truth-tracking conditions yield:

(TT)   S knows that p only if S's belief that p satisfies conditions (T1) and (T2).

The question is whether all items of knowledge conform to (TT).[3] Nozick claims that they do.

There are counterexamples that indicate that the conditions are too strong. Suppose I turn on an electric heater and leave the room. I know that the heater is on in that room. But if the heater were not on, I would still believe that it was. Condition (T1) is not met. On a possible-worlds analysis of subjunctive conditionals, the closest world in which the heater is not on is one in which an overload switch has been triggered or a wire has become disconnected or my wife has switched it off. Events of this sort are unlikely to alter my belief. An adherent of the truth-tracking condition could respond by arguing as follows. Either the world I describe (in which the heater is not on) is very close, in which case I do not have knowledge in the actual because I cannot have knowledge about a heater that, unknown to me, is on the verge of being disconnected. Or the world I describe is very distant in which case there are closer worlds in which I am not oblivious to the fact that the heater is likely to be disconnected, and in those worlds I would not believe that the heater is on.

---

[2] Nozick discusses all examples in terms of a possible-worlds analysis. It does not appear that his account would fare any better under alternative analyses.

[3] The conditions need to be amended so as to fix the method of belief acquisition across the relevant counterfactual situations (Nozick 1981, p. 179), but my criticisms do not depend on this point.

What are the relevant close worlds? I claim that my reading of the conditional is a perfectly natural one. I know that the heater is on, although there are very similar situations that I could be in which the heater is off but in which I continue to believe it is on. It is easy to suspect that our respondent is choosing worlds to fit the theory rather than the other way around. Perhaps I am perversely choosing worlds to counter the theory. This points to a more general problem for the truth-tracking account of knowledge. The truth-values of subjunctive conditionals are simply too indeterminate and too context-dependent for such an account to bear the weight of the task demanded of it, that of distinguishing knowledge from non-knowledge. Our intuitions as to what counts as knowledge may be rough and ready, but our intuitions as to the truth-values of subjunctive conditionals are even rougher and much less ready to make fine distinctions.

However, the context-dependence of subjunctive conditionals may be seen as a virtue of the truth-tracking account because knowledge ascriptions are themselves context-dependent. That knowledge ascriptions are to some extent context-dependent is a reasonable claim. In some contexts (for example, criminal trials or drug-testing), we set a more stringent standard for knowledge than in other contexts. But if the truth-tracking analysis is correct, then variation in ascription of knowledge will not just vary along a stringency scale. Much will depend on what is in the mind of the ascriber and this can vary in multiply different ways. Given that there is considerable agreement in most actual cases, it is doubtful that knowledge is context-dependent to such an extent.

Another objection is that on the truth-tracking account, knowledge is not closed under known implication. Knowledge is closed under known implication if it is the case that if S knows that p and S knows that p implies q then S knows that q. A weaker (and perhaps more plausible) condition has it that knowledge is *closeable* under known implication. In other words, if S knows that p and S knows that p implies q then S can know that q.

The truth-tracking conditions are strong enough to rule out both closure and closeability. Consider:

(a)   I am in my office at the University of Otago
(b)   I am not a brain in a vat in Los Angeles having 'my-office' experiences induced.

According to (TT), I can know (a) but I cannot know (b). I cannot know (b) because if (b) were false I would still believe (b). A brain in a vat undergoing 'my-office' experiences does not believe it is a brain in a vat.

Nozick takes this to be a virtue of his position (1981, pp. 204-17). He agrees with global sceptical claims such as that we cannot know that we are not brains in vats or that we are not the victims of a Cartesian evil demon, but disagrees with narrower sceptical claims that we cannot have knowledge of our actual surroundings. There are grounds for supposing that this is in accord with common sense or intuitive knowledge attributions. Alvin Goldman (1989, pp. 202-04) argues that recent psychological research supports and explains non-closure for cases of global scepticism.

Other apparent anomalies still remain for the brain-in-a-vat example. One is that (TT) denies knowledge of the disjunction of (a) and (b) even though it allows knowledge of one of the disjuncts, namely (a). If the disjunction were true I would believe it, but if it were false, then (b) would be false and I would still believe (a), and hence believe the disjunction (Plantinga 1988, p. 16 & p. 49, fn. 18). Maybe there are grounds for claiming that this outcome accords with a viable notion of knowledge, but it seems unlikely.

The k-causal condition does not rule out closure or closeability. In fact, it deliberately allows for it. However, since it is a necessary but not sufficient condition for knowledge, there is the possibility that there are further conditions on knowledge, which yield non-closure. For example, Goldman's 'relevant-alternatives' condition may do this, as he argues in his (1989). The k-causal condition is open on the question of the closeability of knowledge under known implication (if implication is taken to be logical consequence as defined in Section 5.5) while the truth-tracking conditions rule it out. Given that the issue is controversial, the k-causal condition has the advantage in this respect.

I conclude that the truth-tracking condition on knowledge is either too strong or not clear enough to make the required distinction.

6.3. TRUTH-TRACKING AND PLATONIC OBJECTS

Suppose my objections to the truth-tracking conditions are rejected and those conditions are accepted as providing the best account of the required connection between fact and belief. What implication does this have for knowledge of platonic objects? First, suppose that there are contingent platonic objects. Let **a** be such an object. Although **a** is F, it is possible for it not to be the case that **a** is F, either because it is possible for **a** not to exist or for **a** to be not-F.[4] Consider a situation in which S truly believes that **a** is F.

---

[4] If F is an essential property of **a**, then it will not be possible for **a** to be not-F, but it will still be possible for **a** not to exist.

If the fact that **a** is F is an acausal contingent fact, then it is reasonable to suppose that that fact could change without there being a change in the causal realm. So any close situations in which it is not the case that **a** is F will be situations in which the only changes will be in the platonic realm. That being the case, S would still continue to believe that **a** is F and (T1) will not be met. The truth-tracking conditions rule out the possibility of knowledge of contingent facts about platonic objects.

If we assume, as is usually done, that true mathematical claims are necessarily true, then the antecedent of (T1) will be necessarily false. Believing that there are no satisfactory methods of dealing with such subjunctives, Nozick (1981, p.186) claims that only (T2) is relevant.[5] And given that p is true in all possible worlds, (T2) simply requires that S believe that p in all sufficiently close worlds.

Nozick then argues that if S comes to believe that p on the basis of a method that guarantees truth (such as, according to Nozick, mathematical proof), then S will know that p and (T2) will be met. On the other hand, if S comes to believe that p via a suspect method (such as believing everything his parents tell him), then there will be close worlds in which the method convinces him that not-p and (T2) will not be met (1981, p.186). This claim faces similar objections to those discussed above (with respect to the heater example). Are there not close worlds in which a trivial error in the mathematical proof results in S's not believing that p? Even S's just worrying that an error has been made could have this result. But these objections must be set aside since this section is premised upon the assumption that these sorts of objections can be overcome.

Nozick is making the claim that if someone knows a necessary truth p, then she must have come to believe that p via a method that would result in the same belief in *similar* circumstances. Let us grant him that claim. The method that he supposes can achieve this in the case of mathematical knowledge is mathematical proof. Unfortunately, mathematical proof alone cannot yield knowledge of platonic objects. Mathematical proofs of theorems that assert (or entail) the existence of platonic objects proceed from axioms that assert (or entail) the existence of such objects. Knowledge of axioms cannot be acquired by mathematical proof. Even if it is correct that

---

[5] He could just as easily have claimed that (T1) is vacuously met in such cases, as this would seem to fit with a possible-worlds analysis. If p is necessarily true, then p is true in all possible worlds. So it is not the case that S believes that p in the closest worlds in which p is not true, because there are no such worlds.

(T2) is a necessary (and sufficient[6]) condition for a belief in a necessary truth to be an item of knowledge, this does not make any substantial contribution to the debate on whether knowledge of platonic objects is possible. Nozick admits as much when he states in a footnote that '[t]he topic of how condition [T2] comes to be satisfied for our beliefs about mathematical axioms ... is an interesting one, but is not our topic here.' (p. 685, fn. 23)

The reason that the truth-tracking theory fails to make a useful contribution is that, although it suggests a connection between fact and belief, it does not provide what may be termed a 'metaphysical' connection. Even if we agree that the subjunctive conditional (or something like it) must be met, how it can be met is the crucial issue. Only when we can point to a connection that exists (or could exist) in the world (such as a causal connection) can we be satisfied that knowledge is possible. If platonic objects lack metaphysical connections with human beings then conditions like (T2) cannot be met.

6.4. THE LAW-LIKE CONNECTION

David Armstrong's *reliable indicator* account of (non-inferential) knowledge is based on the model of measuring instruments such as thermometers and barometers (1973, pp. 166-171). A reliable thermometer is one for which the temperature reading is a reliable indicator of the actual temperature. Analogously, beliefs that count as items of knowledge are reliable indicators of actual states of affairs. Armstrong explicates reliable indication in terms of a law-like, or nomological, connection.

> Suppose, on a certain occasion, a thermometer is reliably registering 'T''. There must be some property of the instrument and/or its circumstances such that, if anything has this property, and registers 'T'', it must be the case, as a matter of natural law, that the temperature *is* 'T''. (p. 167)

Applying this idea to the case of knowledge, we have:

(RI)   S knows that *p* only if S is in some condition and/or circumstances such that, as a matter of natural law, if any cogniser is in that condition and/or circumstances, then, if they further believe that *p*, then *p* is the case.

---

[6] Fumerton (1987, p. 179) points out that (T2) is insufficient because someone determined to believe all mathematical propositions presented to him would satisfy (T2) for those which are true.

Putting this more formally, to bring out the role of law-like connections, we have:

(LL)   S knows that p only if S believes that p & $(\exists H)$[S is H & there is a law-like connection in nature such that $(\forall x)$ if x is H, then (if x believes that p, then p)] (p. 168)[7]

Armstrong notes that given that S believes that p, then the truth condition is redundant in his account of knowledge. He argues, as I have, that this is a strength (p. 187).

What are these law-like connections? Armstrong does not give a full account, but he makes three points that he takes to be essential: (1) they are the sort of connections that can, in principle, be investigated by scientific method; (2) the law-like generalisations that record the existence of such connections yield subjunctive conditionals; (3) the connection between belief and state of affairs is an ontological connection but not a *causal* connection (p. 169).

Armstrong's third point is puzzling. After all, in the case of the paradigmatic models, such as thermometers and barometers, the actual physical quantity (temperature, atmospheric pressure) *is* a cause of the reading. It seems that the point he wishes to make is that although it is frequently the case that the fact that p does bring the belief that p into existence, this does not occur in all cases. He does not provide any counterexamples in which beliefs are not caused by their corresponding facts, but does claim that in the case of a reliable watch 'the time of day is not the *cause* of the watch registering the time of day correctly' (p. 175). This is not a particularly convincing example. Even if the time of day and the watch's registering that time are not causally connected, they are k-causally connected. The fact that the time of day is the time that it is will be a logical consequence of any statement of the complex fact that is a cause of the fact that the watch is registering that time. At least, it will be if the watch is not showing that time as a matter of chance.

The possibility of there being a law-like connection between fact and belief that is not a causal connection is suggested by an example of Alvin Goldman (1986, p. 43).[8] Suppose for some particular brain state B that

---

[7] Armstrong restricts p to states of affairs in which some individual **a** has some property F (p. 170), but this will not affect my comments or criticisms.

[8] Goldman offers his example as a counterexample to the sufficiency of Armstrong's account, but that is not relevant here.

whenever a human being believes he is in brain state B, this nomologically implies that he *is* in brain state B. How could such a law-like connection arise? It could arise from a causal connection. The brain state may cause the belief or the belief may cause the brain state or they may both have a common cause. Alternatively, if beliefs are type-identical to brain states, then brain state B may be a constituent of the belief. The relationship could be that of part to whole or even that of identity. But the part-whole relation and the identity relation are not causal. However, we can regard the part-whole and identity relations as degenerate cases of the causal relation since any states so related must have a common cause. As such, the belief and the brain state are k-causally connected.

I conclude that there is no reason to believe that the reliable indicator condition can be nomologically met unless the k-causal condition is also met. That being so, if (LL) is necessary for knowledge, then so is the k-causal condition (SC*).

6.5. LAW-LIKE CONNECTION AND PLATONIC OBJECTS

Suppose I am wrong and that it is possible for there to be a law-like connection between fact and belief in the absence of a k-causal connection. Could this possibility account for knowledge of platonic objects? Armstrong does not think so. He does not believe that there are any entities which lack causal powers or, if there were, that we could have knowledge of them. He believes that mathematical knowledge, for example, is knowledge of certain kinds of universals (properties and relations), but although he is a realist about universals, he does not believe that they exist outside the causal nexus (1973, p. 102; 1978a, pp. 130-32). Further, he believes that only spatio-temporal particulars can stand in law-like connections with belief states, so the only objects we can have existential knowledge of are concrete objects (1973, p. 170). Finally, his account of knowledge of general facts is in terms of knowledge of particular facts, so even the possibility of general platonic knowledge is ruled out (1973, p. 90).

Armstrong's account of law-like connections is not detailed enough for us to tell readily whether such an account could be expanded to show how platonic objects as well as concrete objects could come to be nomologically connected to our beliefs. There is a difficulty for such an expansion (Casullo 1992, pp. 566-67). Armstrong's second requirement for law-like connections is that the law-like generalisations that describe them should support subjunctive conditionals. In the case of knowledge, these must be conditionals such as:

(C)   If not p, then it would not have been the case that S believed that p.

But this is Nozick's (T1). I argued in Section 6.3 that if p is a contingent platonic truth then (C) is unlikely to be met. On the other hand, if p is a necessary platonic truth, then (C) will be a subjunctive conditional with an impossible antecedent. In the absence of a satisfactory account of the truth conditions for such conditionals it is doubtful whether statements like (C) have a determinate sense. If (C) is vacuously true, then so is:

(C′)   If not p, then it *would* have been the case that S believed that p.

If this is the case, then (c) is not of any epistemic interest. Given that Armstrong does not admit the possibility of platonic knowledge, it is not surprising that his condition on knowledge does not allow for such knowledge. Nor is it surprising that it is difficult to see how his account could be modified to accommodate platonic knowledge.

Armstrong's reliable-indicator theory is a reliabilist theory of knowledge. Another version of reliabilism is process reliabilism. The reliable-process condition, which requires that for a belief to be known it must be caused by a reliable process, does not entail the truth of the belief. A reliable process can cause a false belief, so long as the process is not perfectly reliable. In Section 4.9, I showed how process reliabilism may be 'gettierised'. I shall not discuss process reliabilism in this chapter, which is devoted to 'connectionist' theories. However, I discuss it further in Chapter 8, where it raises some interesting issues of relevance there.

6.6. THE EXPLANATORY CONNECTION

Alan (A.H.) Goldman suggests that the important condition for knowledge is that the fact that p must explain, and render (significantly) more probable than it otherwise would be, the belief that p (1988, pp. 19-37). He further suggests that a necessary condition for this connection to hold is that the probability that S believes that p given that p is true should be significantly greater than the unconditional probability that S believes that p. So we have:

(E1)   S knows that p only if the fact that p explains S's belief that p,

or

(E2)   S knows that p only if pr(S believes that p /p) >> pr(S believes that p).

The idea is that in the usual case of my phoning my aunt in Eketahuna, the fact that it is raining will play a significant role in an explanation of my belief that it is raining. Furthermore, in the circumstances, I am much more likely to believe that it is raining if it *is* raining than I am to believe it regardless of the weather. However, in the 'gettierised' version in which my aunt is confused and her weather report is correct only by chance, the fact that it is raining does not play a role in explaining my belief, and the fact that it is raining does not, in the circumstances, increase the likelihood of my having that belief.

One kind of possible counterexample to this account can be readily dealt with. When I am looking at the fire, the fact that the smoke is rising from the chimney does not explain, or make more probable, my belief that it is, and yet I know that the smoke is rising. But we can patch up the theory along the same lines as we did for the causal theory; that is, we can weaken the requirement so that the fact and belief may be connected via some other fact that explains both the fact and my belief.

The main problem with this condition is the difficulty in determining if it has been met. Does the fact that all protons have a rest mass of 1.672614×10^{-24} grams explain my belief that they do, any better (or to any greater extent) than the fact that all observed protons have such a mass and the fact that I assume an appropriate inductive principle? Does the universal fact make my belief any more probable than it would otherwise be? It certainly does not make it significantly more probable than it would be if a few distant (or even a few close but unmeasured) protons had a different mass.

Goldman suggests that the relevant probabilities can be calculated by ranging over close possible worlds. He states that:

> The probability of a proposition's being true is measured by the number of close or accessible worlds in which it is true (or by the ratio of those in which it is true to those in which it is false). (1988, p. 26)

This cannot be right. Probability is measured by a number between zero and one. Both of his suggested measures can readily yield a number greater than one. But from his subsequent remarks, I conclude that he means that the probability of A is the proportion of close possible worlds in which A is true, while the probability of A given B is the proportion of close possible worlds in which both A and B are true out of those in which B is true.[9]

That only close worlds are relevant Goldman sees as an advantage, particularly in comparison to the truth-tracking condition. He gives the following example. A father is playing tennis with his young son when he hears from a radio broadcast that an assassination has occurred at a distant location. Thus he knows that his son is not the assassin. But suppose that if his son had been guilty, the father would still not have believed that he was. His love for his son would, in the circumstances, drive him to distraction. The truth-tracking condition is not met but, according to Goldman, the explanatory condition is because facts that explain the son's innocence also explain the father's belief in his innocence. What is puzzling is that the probability condition does not appear to have been met. In all close possible worlds the son is innocent. After all, he is only ten years old, a great distance from the scene of the crime, and lacks the motivation, temperament, and skills of an assassin. It follows that on Goldman's method of calculating the probabilities, the probability that the father believes that his son is innocent given that he is will be identical to the antecedent probability that he so believes.

A similar problem arises if we suppose that the belief concerns a law of nature. If we make the reasonable assumption that close worlds are those with the same laws of nature as the actual world, then pr(S believes that p /p) will equal pr(S believes that p). Defenders of the proposal will no doubt object that we must obviously range over possible worlds in which p is false, but then we are in danger of choosing our possible worlds in order that our calculations meet our intuitions. As with the truth-tracking analysis, we might suppose that these problems arise because knowledge is highly context-dependent. If so, my previous remarks in this regard will apply. My conclusion is the same as that for the truth-tracking account. Either the explanatory condition is too strong or it is too unclear.

6.7. EXPLANATION AND PLATONIC OBJECTS

Let us suppose, as we did with the truth-tracking account, that these problems can be overcome. It can be readily seen that the explanatory connection as explicated by the probability condition will not yield knowledge of platonic objects. Let p be a contingent fact concerning a contingent platonic object and suppose that S actually believes that p. Unless we have some

---

[9] He states that 'if B raises the probability of A, then the ratio of close worlds in which A obtains is higher in the set of worlds in which B obtains than in the entire set of worlds' (p.26)

account of how the truth-value of this acausal fact could make a difference to the probability of S's believing that p, we have no reason to suppose that the fact that p could raise the probability that S believes that p.

Let p be a necessary truth concerning platonic objects and suppose that S believes that p. Since p is true in all possible worlds, it will be true in all close worlds, and pr(S believes that p/p) will be identical to pr(S believes that p).

Perhaps Goldman's analysis of probability or the probability condition itself is a mistake. It does not follow that that his explanatory condition (E1) is also mistaken. Recall:

(E1)   S knows that p only if the fact that p explains S's belief that p.

What would count as a good explanation of the fact that S has the belief that p? Since the fact to be explained appears to be a singular physical fact, the most likely candidate for a genuine explanation is a causal explanation. If that is right, then the only way a fact can play a significant role in an explanation of a belief in that fact is if the fact is causally connected to the belief.[10] In other words, a clearly explicated explanatory condition on knowledge reduces to a causal condition.

It should be noted that Goldman may intend his analysis of knowledge to apply only to empirical knowledge. But this is by no means clear. Although the title of his book is *Empirical Knowledge* and he observes that '... of course empirical knowledge differs from mathematical knowledge, ethical knowledge, and other types', he also claims that '[an] analysis should capture what is common across contexts and types while allowing for variation in application' (1988, pp. 21-22). Perhaps he does assume that his analysis throws some light on mathematical knowledge. Not that Goldman's intentions are strictly relevant. We are interested in whether or not his account does actually shed light on the problem of knowledge of platonic objects.

The important question is whether an acausal fact can play a role in explaining someone's belief in that fact. In the absence of a plausible mechanism, we must remain sceptical of platonic knowledge, even if we accept that Goldman's account is along the right lines. In a footnote, he states that '...the truth of [a mathematical] theorem helps to explain the possibility of proofs for it, and the proofs help to explain the beliefs of the

---

[10] David-Hillel Ruben (1990) argues that non-causal, but metaphysically real, relations (in particular identity and part-whole relations) may be cited in an explanation (pp.209-33). But, in the context of knowledge conditions for beliefs, causal theorists will accept that causal connections may be extended by such relations. See also my Section 6.4.

mathematicians' (p. 390). But an appeal to mathematical proofs faces the same difficulties as it did for the truth-tracking account. Knowledge of a proof cannot explain beliefs in axioms, and it is knowledge of axioms that is crucial.

6.8. THE NONDEFEASIBILITY CONDITION

Nondefeasibility accounts of knowledge provide another kind of truth-entailing condition on beliefs. A simple version requires that:

(D)   S knows that p only if there is no truth q such that if S believed that q, then S would no longer be justified in believing that p.[11]

If p were not true, then $\sim p$ would be a truth such that if S believed it, then S would no longer be justified in believing that p. (D) is a condition which can only be met if p is true.

(D) involves a subjunctive conditional and thus faces the usual problem of evaluating the truth condition of that conditional in hypothetical situations. The condition is also much too strong. It can be readily shown that there is always a truth that will do the job of rendering the belief p unjustified. Let r be the true proposition saying what numbers will win the next Lotto. Then $\sim p \vee r$ will be such a truth. If S came to believe $\sim p \vee r$, then, since r is highly improbable, S would be justified in believing $\sim p$. S would no longer be justified in believing that p (Harman 1973, p. 152).

When attempts are made to capture the original intuition underlying the nondefeasibility, while avoiding such counterexamples, nondefeasibility accounts rapidly become complex. (See, for example, Lehrer (1974, p. 223-24; 1990, pp. 147-49). This alone is a good reason to doubt that such an account explicates the concept of knowledge. The complexities usually involve counterfactuals in which we suppose that S has acquired all (relevant) true beliefs and then we must decide if she still has her original justification. Once again, intuitions as to the truth of such subjunctive conditionals will not bear the weight put on them.

---

[11] Adapted from Pappas & Swain (1978, pp. 27-28). See also Lehrer & Paxson (1969).

6.9. NONDEFEASIBILITY AND PLATONIC OBJECTS

Let us set aside these problems for a nondefeasibility condition on knowledge and consider whether such a condition allows the possibility of platonic knowledge. This is not easily decided. The kinds of facts that might be cited as defeating justification are controversial. Suppose S believes the platonic fact that p and his belief is highly justified, say by the authority of the mathematical community. Would not S's justification be defeated by his coming to believe that the objects of his belief have no causal powers and that human beings are wholly physical beings? (S might have read this book.) But the relevance of these facts, if they are facts, to the possibility of our having platonic knowledge is just what is in dispute. We cannot know if beliefs about platonic objects are nondefeasible unless we know whether we can be justified in believing that we can have platonic knowledge. The primary aim of this book is to argue that we are not so justified.

6.10. CONNECTIONIST ACCOUNTS AND PLATONIC KNOWLEDGE

The truth-tracking and nondefeasibility accounts depend on subjunctive conditionals whose truth-values are too indeterminate to provide a satisfactory condition on knowledge. The explanatory account depends on a notion of conditional probability, which is similarly indeterminate. Attempts to provide a precise reading of those accounts yields either a condition that is too strong for knowledge or a connection that depends on their being a causal link between fact and belief. The law-likeness account is also dependent on a causal link. None of these accounts, when given a precise and plausible reading, allows for the possibility of platonic knowledge.

CHAPTER 7

EXISTENCE CLAIMS AND CAUSALITY

7.1. EXISTENTIAL PLATONIC KNOWLEDGE

I have argued that a connection between fact and belief that guarantees truth is a necessary condition for knowledge. I have further argued that the connection must be causal in nature, at least in the attenuated sense of being a k-causal connection. If my claim is correct, then we do not have knowledge of platonic objects. I have discussed difficulties for sustaining these claims in the face of various objections and counterexamples. In this chapter, I argue that a global causal requirement is not the only basis for a causal objection to platonism. By a 'global' requirement, I mean one that applies to all knowledge. My challenge to the platonist will be to ask how we can know that platonic objects, as such, exist. I shift the focus of the debate away from causal constraints on knowledge in general and towards causal constraints on *existential* knowledge, by which I mean, knowledge that certain entities exist.[1]

7.2. HALE'S ARGUMENT

Bob Hale defines a strong causal theory as one that requires a causal connection between fact and belief, as distinct from *weak* causal theories, which require only that beliefs be caused in an appropriate way (1987, p. 93). He argues that no strong causal theory is acceptable and concludes that there are no serious causal objections to platonic knowledge. From Hale's

---

[1] This chapter is based on Cheyne (1998).

somewhat labyrinthine treatment of the causal objections to platonism (1987, pp. 92-101), we can distil the following argument:

(H1)   The only way a causal theory can block knowledge of platonic objects is by requiring that every known fact must be causally connected to the knower's belief in that fact,

(H2)   Some known facts, in particular general (universal and existential) facts, need not (perhaps cannot) be causally connected to their knower's beliefs,[2]

therefore,

(H3)   Any causal theory that blocks knowledge of platonic objects will also block knowledge of general facts and hence be false,

therefore,

(H4)   There can be no effective causal objection to knowledge of platonic objects.

Hale does not dispute that often a fact is connected to the knower's belief in that fact. What he does dispute (in H2) is that *all* known facts are so connected. He further claims that no theory can rule out platonism unless that theory requires such a connection. *Prima facie*, he is correct. If not all known facts need to be causally connected to the knower's beliefs, why shouldn't some of those unconnected but known facts be the disputed 'facts' of platonism?

7.3. CAUSAL CONSTRAINTS ON EXISTENTIAL KNOWLEDGE

I argue that even if such causal links are not always necessary for knowledge, there are other causal constraints that are necessary for existential claims, and that platonic objects cannot meet such constraints. In other words, I reject Hale's first premise and offer a causal theory that blocks knowledge of platonic objects but does not meet the requirement of (H1). It might be argued that my theory is not a genuine causal theory of knowledge. In that case, I would claim that Hale's argument is not valid; (H4) does not

---

[2] Here I follow Hale's practice of referring to both universal and existential facts as 'general' facts, in contrast to 'singular' facts.

follow from (H3), since there can be an effective causal objection which is not based on a causal theory of knowledge.

I wish to focus the debate on the platonists' claim that we can know that platonic objects exist. I do this by asking how we know that certain kinds of objects exist. I propose an answer to this question which, if correct, suggests that we cannot know that platonic objects exist.

My argument is as follows:

(CE)   We cannot know that F's exist unless our belief in their existence is caused by at least one event in which an F participates,

(C2)   Platonic objects cannot participate in events,

therefore,

(C3)   We cannot know that platonic objects exist.

More needs to be said about (C2). I am using 'participate' in a technical sense. An object participates in an event if its presence makes a difference to the causal powers of that event. Platonic objects, since they lack causal powers, cannot participate in events in this sense. It is perhaps theoretically possible for acausal objects to be present in an event without participating in it. Suppose there are acausal objects with spatio-temporal locations. Let us call them 'sprites'. Suppose there is a sprite in my coffee cup. That sprite may be present in (or even part of) the event of my picking up my coffee cup, but it does not participate in that event. Even if, for some bizarre reason, my picking up my coffee cup causes S to believe that there is a sprite in my coffee cup, it is not the case that an event in which a sprite participates causes S's belief.

7.4. ABOUTNESS AND EXISTENCE

My claim is not that we cannot know anything about certain objects if we are causally isolated from all events in which they participate—only that we cannot know that objects of that kind exist. In 1871, Mendeleeff knew many salient facts about germanium, although germanium itself was an unknown element preceding arsenic in his Periodic Table. For example, he knew that germanium, if it existed, would have an atomic weight of about 72, a density of 5.5 gm/cm^3, and form an organo-metallic compound $Ge(C_2H_5)_4$ that boils at 160°C. He inferred those facts from the Periodic Table. But he did not know that germanium existed until Winckler discovered the metal in 1887 (Holton & Roller 1958, p. 423). What Mendeleeff knew in 1871, he could have known if this planet had been devoid of germanium atoms. If he believed, prior to 1887, in the existence of germanium, that belief, although true, would not count as knowledge. It could only be a lucky guess, unless it was actually caused, in an appropriate way, by events in which germanium atoms participated. The proviso 'in an appropriate way' must be added because it is possible that an event involving germanium could give rise to a lucky guess without that belief being an item of knowledge. However, it is not necessary to spell out what that appropriateness involves. Objects that cannot participate in events cannot participate in 'appropriate' events.

As with germanium, I do not dispute that there is a sense in which we could know many things about various kinds of platonic objects. But that is not enough for platonism. Platonism requires that we know that platonic objects exist, otherwise it is no more than a version of postulationism or 'if-thenism'. If-thenist theories attempt to avoid ontological commitment to platonic objects by claiming that all mathematical statements should be construed as conditional statements (see Section 5.5). There are difficulties in formulating a plausible version of if-thenism, but the fact that some philosophers of mathematics have made strenuous efforts to do so stems from a clear desire to avoid existential claims. These efforts are discussed further in the next section.

My dispute is with platonism, not if-thenism, and we cannot have a genuine platonist theory without ontological commitment to platonic objects. It is not enough for the platonists to claim that we can have knowledge without causal connection. They must claim that we can have *existential* knowledge without causal connection.

7.5. THE 'NON-EXISTENTIAL' TRADITION

There is a tradition in the philosophy of mathematics that seeks to avoid existence claims without abandoning the notion that we have mathematical knowledge. Although Descartes did not eschew all mathematical or abstract objects, he thought that 'Arithmetic, Geometry and other sciences of that kind ... only treat of things that are very simple and very general, without taking great trouble to ascertain whether they are existent or not', and that it is because of this that they 'contain some measure of certainty and an element of the indubitable' (1641/1969, 'Meditation I' p. 168). He notes that we can have geometrical knowledge of a chiliagon (a one thousand-sided polygon) without being able to imagine one clearly, let alone see one ('Meditation VI' p. 209). It would seem to follow from this that we could have such knowledge without such a figure ever existing, although Descartes does not draw this conclusion.

Locke argues that (apart from our knowledge of the existence of our-selves and of God) we can only have knowledge of the 'real existence' (that is, existence independent of the mental) of particular things via sensation. Knowledge by sensation is less certain than knowledge by intuition or demonstration, but knowledge by intuition or demonstration can only pro-vide knowledge of the agreement or disagreement between ideas. Mathe-matical knowledge, being certain, can only be, according to Locke, know-ledge of the relations between ideas and not knowledge that involves the existence of particular things. For Locke, the abstract objects of mathematics exist only as ideas in the mind or perhaps as the relations between ideas. They cannot have independent existence, as do the abstract objects postulated by platonists (1690/1976, Book IV, esp. ch. 1, 2, 9, 11).

Bertrand Russell believes that mathematical knowledge must be knowledge of necessary truths. '[P]ure mathematics,' he writes, 'aims at being true ... in all possible worlds, not only in this higgledy-piggledy job-lot of a world in which chance has imprisoned us.' (1919, p. 192) He also believes that all existence is contingent. 'There does not even seem any logical necessity why there should be even one individual.' (p. 203) He concludes that mathematical knowledge cannot include existence claims and proposes an if-thenist construal of mathematics (p. 204).

The arguments of Descartes, Locke and Russell have a similar form. Each identifies a certain property of existential knowledge (dubitable, empir-ical or contingent), points out that mathematical knowledge does not have that property, and concludes that mathematical knowledge cannot be exist-ential knowledge. I employ a similar argument, not to demonstrate that mathematical knowledge is non-existential, but rather as a *reductio* on platonism. Existential knowledge has a certain property (causal connection with the objects of that knowledge), platonic knowledge must lack that property, therefore, platonic knowledge is non-existential. But knowledge of the existence of platonic objects is essential to platonism. Therefore, platon-ism is false.

7.6. DEFENDING THE EXISTENTIAL CAUSAL CONDITION

My argument in this chapter depends on a defence of the causal condition on existential knowledge:

(CE)   We cannot know that F's exist unless our belief in their existence is caused by at least one event in which an F participates.

This condition does not face many of the difficulties faced by the global causal condition:

(SC)   S knows that p only if the fact that p is causally connected to S's belief that p.

Because (CE) refers specifically to event causation, there are no problems with the notion of fact causation. This also means that there are no problems with citing universal or existential facts as causes. Indeed none of the problems involving knowledge of universal facts arises because such facts do not entail the existence of any objects. There is no need to extend causal links with logical connections, so there is no problem with necessary truths (even necessary existential truths, if such there be) meeting the condition vacuously (Section 5.5). The problem of establishing a causal connection with non-local quantum facts is avoided. The EPR set-up does not allow us to acquire knowledge of the existence of previously unknown kinds of objects, although it may allow us to acquire new knowledge about objects whose existence we have come to know of by causal contact (Section 5.9). In the 'beheading' counterexample, there are no objects whose existence is believed in without that belief's being caused by an event in which such an object participated (Section 5.7). It is not possible to devise examples along similar lines in which condition (CE) is not met, so it is not necessary to allow causal overdetermination as part of an appropriate causal connection. There are, however, other possible counterexamples, which I consider in Section 7.10.

7.7. THE CASE FOR AN EXISTENTIAL CAUSAL CONDITION

In Section 4.2, I discussed an argument in defence of (SC) which considers the sort of reasons that we offer when we wish to reject a claim that S knows that p. Even if we are satisfied that p is true, that S believes that p, that there is nothing wrong with S's reasoning powers, and so on, it is still legitimate to argue that S does not know that p because S could not have made causal contact with the fact that p. There are difficulties with sustaining this claim for any p (Chapter 5). However, in the case of a claim to know that F's exist, the argument that S has not had, or could not (in the circumstances) have had, causal contact with an F is more clearly a strong ground for rejecting the knowledge claim. The example of Mendeleeff and germanium reinforces this point. In 1871, Mendeleeff may have believed on the basis of the gap in his periodic table that germanium might exist and we may even agree that it was a reasonable assumption, but it still remained for the element to be discovered. Knowledge that germanium exists required causal contact, however remote, with atoms of germanium.

A more recent example comes from the announcement of the discovery of the top quark at Fermilab, the high-energy physics laboratory near Chicago (Close & Maddox 1994). Quarks are the basic ingredients of nuclear particles such as protons and neutrons. All the other kinds of quarks predicted by the current standard theory of particle physics (up, down, charm, etc.) had been detected previously. So scientists were confident that the top quark which completes the picture must exist. The announcement of its discovery came when scientists at the laboratory believed they had detected the effects of the manufacture and subsequent decay of these particles.

> But why all the excitement about the top quark, of whose ultimate existence nobody has much doubt? It has something to do with the principle that seeing is believing; inference, however neat the underlying theory, is always less persuasive than demonstration. But it would also be galling for high-energy physicists to be basing calculations of the properties of particles on the assumption that the top quark exists when they had not been able to manufacture a single specimen. (Close & Maddox 1994, p. 805)

There is a suggestion here that perhaps scientists already knew by inference that top quarks exist and that the detection simply increased their confidence. But later in the article, it becomes clear that that inference was also based on events in which the top quarks participated.

> [T]here is already good evidence for the existence of the top quark... [It] has also already manifested itself in its perturbation of precisely observable quantities at CERN's LEP... [T]he bosons Z^0 and $W^{\pm}$ were similarly inferred from precision data at CERN.[3] (p. 805)

---

[3] CERN is the European Organisation for Nuclear Research. LEP is their Large Electron-Positron collider.

The history of science is replete with examples of entities being posited and followed by attempts to detect the existence of those entities, where detection involves the investigator's being causally connected to events in which the posited entities participate. Only then can it be announced that it is known that the entities exist. Ian Hacking (1983) discusses a number of cases, in particular the 'philosopher's favourite entity...the electron' (p. 262). After discussing the work of Thomson and Millikan, he details the building and successful use of a polarising electron gun (PEGGY II). By observing the effects of the scattering of polarised electrons scientists detected the non-conservation of parity. Hacking concludes that:

> The 'direct' proof of electrons and the like is our ability to manipulate them using well-understood low-level causal properties... The best kinds of evidence for the reality of a postulated or inferred entity is [sic] that we can begin to measure it or otherwise understand its causal powers. The best evidence, in turn, that we have this kind of understanding is that we can set out, from scratch, to build machines that will work fairly reliably, taking advantage of this or that causal nexus. (p. 274)

Hacking is concerned with rejecting the anti-realist claim that we cannot know of the existence of hypothetical entities if we cannot 'directly' observe them. His response is that we can come to know of their existence when we interact with them, in particular when we manipulate them to 'interfere in other more hypothetical parts of nature' (p. 265). He backs up this claim by noting that scientists do become realists about such entities following successful experiments of that kind.

Another interesting example is the discovery of the planet Neptune in 1846. This discovery has been rightly described as 'a noble triumph of theory' and 'one of the greatest triumphs of theoretical astronomy' (Grosser 1962, pp. 120 & 121). The triumph consisted in the use of theory, not only to postulate the existence of the planet, but also to predict its position and size with such accuracy that its existence could be confirmed by more direct observation.

In 1845, the young English astronomer John Adams wrote:

> According to my calculations, the observed irregularities in the motion of Uranus may be accounted for by supposing the existence of an exterior planet, the mass and orbit of which are as follows... (Grosser 1962, p. 88)

In 1846, the French mathematician Urbain Leverrier wrote that:

> it is impossible to satisfy the observations of Uranus without introducing the action of a new Planet, thus far *unknown*... The actual position of this body shows that we are now, and will be for several months, in a favourable situation for the *discovery*. (*Ibid.*, p. 116, my emphasis.)

Using Leverrier's calculations, the German astronomers Galle and d'Arrest observed the planet Neptune by telescope on 23 September 1846. Galle triumphantly wrote to Leverrier, 'The planet whose position you have pointed out *actually exists*.' (*Ibid.*, p. 116, his emphasis.)

Note that it is not unreasonable to claim that Adams and Leverrier knew quite a bit about Neptune before they knew of its existence. Perhaps they also knew of its existence before the sighting, even if they were not prepared to claim as much. Herschel appears to have thought so. Speaking of the postulated planet in early September 1846, he said:

> We see it as Columbus saw America from the shores of Spain. Its movements have been felt, trembling along the far-reaching line of our analysis with a certainty hardly inferior to ocular demonstration. (*Ibid.*, p. 114)

Herschel's analogy is distinctly odd. Columbus had no idea that there was a large landmass between Spain and the Indies. Even when he set foot on it, it is doubtful that he realised this. If he had postulated its existence before departing the shores of Spain, he would have been in a similar position to those who later postulated the existence of a southern continent. Their only ground was a desire to fill an empty space, a desire for symmetry. Even if there had been a large southern continent, this ground was inadequate for a knowledge claim. The nineteenth-century astronomers were in quite a different position. Theoretical considerations played a major role in their speculations, but causal interaction was crucial. It was the motion of the unseen planet that caused the perturbations in the orbit of Uranus that they sought to explain by their postulation. If that indirect interaction, together with their analysis, was sufficient for their knowing of Neptune's existence, so be it. It was still necessary that events involving Neptune caused their belief. No movements, or any other events, involving America gave rise to Columbus's postulation (assuming that he made one).

Scientists are not content just to 'save the phenomena'. They continually strive to discover the existence of the entities that they postulate. They only claim to have done this when they have interacted with them in some way. The best explanation for this behaviour is that it stems from their conviction that they do not know that the entities exist unless they have interacted with them. 'Seeing is believing' is not literally true. A better slogan might be 'Interacting is knowing'.

Another argument for an existential causal condition stems from the apparent incompatibility of platonism with some of our best-confirmed empirical laws. W.D. Hart argues for the incompatibility of platonism and a naturalised epistemology. His argument doesn't apply simply to existential knowledge, but it does apply solely to knowledge that has existential knowledge as a component.

> Granted just conservation of energy, ...when you learn something about an object, there is a change in you...[which] can be accounted for only by some sort of transmission of energy from, ultimately, your environment to...your brain. And I do not see how what you learned can be *about* that object (rather than some other) unless at least part of the energy that changed your state came from that object. (Hart 1977, p. 125)

It is clear that Hart is using 'about' in a sense which means that 'I know about that object' entails the existence of the object concerned. If not, then he is mistaken, because, as I have noted, there is a sense in which we can know about non-existent objects. (One thing I know about fairies is that they don't exist.) So I take him to be saying that 'when you learn of an object, *inter alia*, that it exists, then there is a change in you, ...' If he is right, then it follows that we cannot know that a particular object exists unless we are causally connected with an event in which that object participates. Learning something 'about' a particular object in this sense entails learning that at least one object of a certain kind exists, and the existential causal condition (CE) follows.

7.8. HALE'S RESPONSE

As noted above, Hale's argument relies on the premise that only a strong and global causal condition on knowledge can threaten platonism. Hale considers the possibility that anti-platonists might agree that a strong causal theory cannot accommodate knowledge of general (universal and existential) empirical truths but respond by attenuating their theory so that it applies only to singular facts. Such anti-platonists would, according to Hale, need 'to explain and justify [their] differential treatment' (1987, p. 97). He suggests that anti-platonists might motivate such differential treatment by claiming that:

> ...while there is reason to dilute the causal theory to accommodate knowledge of general (empirical) truths, no reason has been found for supposing that it is not, quite generally, a condition on knowing a singular truth that the truth-conferring fact shall be causally related in an appropriate way to the subject's belief — this account works well...outside mathematics, so special grounds are needed for rejecting the requirement there. (p. 97)

I have been arguing that we should treat existential claims differently from non-existential claims, rather than treat singular claims differently from general claims. I also require only that the subject be causally connected with any event involving one of the objects of knowledge, rather than with the fact that is believed. But, the claim in the quotation is, *mutatis mutandis*, close enough to the one I have been making.

Hale attacks this claim by arguing that our mathematical knowledge differs in relevant ways from empirical knowledge. He says that knowledge of general empirical truths depends upon knowledge of singular ones, while for number-theoretic knowledge, for example, the dependence may flow in the opposite direction. 'Knowledge of the most general and fundamental truths of...number theory, coupled with mastery of the necessary means of deduction, suffices, at least in principle, for knowledge of any singular number-theoretic truth.' (p. 98)

This is false. If number-theoretic knowledge is obtained by deduction, then it is obtained by deduction from axioms. If those axioms are the Peano axioms, then they contain the singular statement '0 is a number'. Any alternative set of axioms from which number theory can be deduced will also contain at least one singular statement (or statement from which a singular statement may be deduced). Now it may be that a characteristic of mathematical knowledge is that many singular truths of number theory may be deduced from a number of general axioms plus only one singular truth, while empirical knowledge is characterised by the (non-deductive) inference of general empirical truths from a large number of singular truths.[4] But this gives no grounds for supposing that differing conditions hold in each case for the acquisition of the *non-inferred* singular facts. And knowledge of such facts is necessary in both cases, if the process is to get started.

Perhaps I have been overlooking the fact that by general facts, Hale means existential facts as well as universal facts. The singular fact '0 is a number' can be reformulated as the non-singular fact 'There exists something which is zeroish', where to be zeroish is to have the properties of the (unique) first member of the natural numbers. Perhaps Hale's claim is (or should be) that the number-theoretic truths can be deduced from this existential fact plus the other general facts of the Peano axioms. Since Hale is supposing that causal connection with general facts (in his sense of general) is not necessary for knowledge, then his argument goes through. But now the distinction I make between knowledge of existential and non-existential facts becomes crucial. Number theory cannot be deduced from axioms that con-

---

[4] Deductivists will claim that both are cases of deducing truths from singular and general axioms. It is just that in the empirical case, the general axioms are usually suppressed. See Musgrave (1989).

tain no existential claims (or imply no such claims). The existential causal condition (CE) works well outside mathematics and we still lack special grounds for rejecting it within mathematics.

7.9. THE NECESSARY EXISTENCE OF PLATONIC OBJECTS

Faced with the challenge of providing grounds for rejecting (CE) in the case of mathematics or some other platonic domain, a common platonist response is to suggest that what distinguishes platonic truths from non-platonic truths is that they are *necessary* truths. They argue that it is reasonable to suppose that such truths can be known by some sort of *a priori* process, one which does not require any (or at least any particular kind of) causal interaction with the external world.[5] I already appear to have conceded as much with my notion of k-causal connection. Beliefs in necessary truths which are a logical consequence of *any* proposition vacuously meet the k-causal condition in the sense that all that is required is that they be caused by some fact. The *content* of the causing fact is irrelevant to their meeting the condition. But I do not concede it in the case of existential facts. My notion of logical consequence blocks vacuity in such cases.

A problem for platonists who claim that it is the necessity of platonic truths that distinguishes them epistemically from non-platonic truths is that it requires that platonic entities not only exist but exist necessarily. For example, those who are platonists about numbers claim that 'There is a prime number greater than 1000' is a necessary truth, which entails that 'There are numbers' or 'Numbers exist' is a necessary truth. Now this cannot be logical necessity. I have already noted Russell's objection to the notion that any entity should exist as a matter of logical necessity.

Russell's objection can be given extra force by showing that contradictions arise from the notion of necessary existence, when that necessity is taken to be logical necessity.[6]

First of all, the notion allows us to argue from the logical possibility of necessary entities to their actual existence. For example, it is logically possible that numbers exist, or in possible-worlds terms, there is a possible world in which numbers exist. On a platonist construal, it is a conceptual truth about numbers that if they exist then they exist necessarily. Then in the possible world in which numbers exist, they exist necessarily. But that is to

---

[5] Hale (1987, p. 101) suggests this response.
[6] The following arguments draw on conversations with Charles Pigden and unpublished writings by Charles Pigden and Rebecca Entwisle.

say that they exist in all possible worlds. Since the actual world is a possible world, then numbers exist in the actual world.[7]

Formally, the argument is:

(N1)	$\lozenge (\exists x)Nx$	(Possibly numbers exist)
(N2)	$\square ((\exists x)Nx \supset \square (\exists x)Nx)$	(Necessarily, if numbers exist then necessarily numbers exist)
$\therefore$ (N3)	$\square (\exists x)Nx$	(Necessarily numbers exist)
$\therefore$ (N4)	$(\exists x)Nx$	(Actually numbers exist).

Note that this argument is only valid in a strong modal logic such as S_5. It is not valid in the weaker S_4, for example.[8]

Platonists might respond that this argument is not so much a problem for their position as a vindication of it. But surely, even the most ardent platonist should not think it so easy to prove the existence of platonic objects. Besides, if the above argument is sound, then so is the following parallel argument for the existence of number-excluders. (Number-excluders are entities that exist necessarily, if they exist at all, and whose existence in a world excludes the existence of numbers in that world.)

(E1)	$\lozenge (\exists x)Ex$	(Possibly number-excluders exist)
(E2)	$\square ((\exists x)Ex \supset \square (\exists x)Ex)$	(Necessarily, if number-excluders exist then necessarily number-excluders exist)
$\therefore$ (E3)	$\square (\exists x)Ex$	(Necessarily number-excluders exist)
$\therefore$ (E4)	$(\exists x)Ex$	(Actually number-excluders exist)

Since (N4) and (E4) cannot both be true, both arguments cannot be sound.

---

[7] This argument is a variation on Descartes' ontological argument for the existence of God, (164/1969, 'Meditation V' pp. 204-05). The arguments of Malcolm (1960) and Plantinga (1974, ch. 10) are, in essence, variations on Descartes' argument.

[8] See Hughes & Cresswell (1968, ch. 4).

There might be some queasiness concerning such weird entities as number-excluders, although it is difficult to see on what grounds someone could insist that they are a logical impossibility whilst numbers are not. But another contradiction is even more readily at hand. Consider the possibility that numbers do not exist:

(N') $\Diamond \sim (\exists x)Nx$

(N3) is logically equivalent to the negation of (N'):

(N3$^{\equiv}$) $\sim \Diamond \sim (\exists x)Nx$

Now (N3), and hence (N3$^{\equiv}$), follows from (N1) and (N2). So, (N1) and (N') cannot both be true, at least if (N2) is true. We might resolve this difficulty by agreeing that either (N1) or (N') is true, but not both. But further difficulties arise when we want to discuss which platonic objects exist. There seem to be many possibilities. Sets exist but numbers do not, or real numbers exist but not imaginary numbers, and so on. And then there are questions as to whether the natural numbers are identical to the Peano numbers or the Fregean numbers, or do both kinds exist? Further complications arise with the possibility of non-platonic but necessary beings such as God. Do both numbers and God exist, or only one, or neither?

Now, it is natural to speak (as I have done) of these as different possibilities. But if the above suggestion is accepted, then this must be a mistake. Only one of the many (possible?) combinations would be a genuine possibility. All the rest must be logical impossibilities! That just seems wrong. If there is no logical contradiction in the notion of sets or of the natural numbers or of a supreme being, then we want to be able say that it is logically possible that they exist without being committed to their actual existence. The trouble with entities with necessary existence is that they spread through all of logical space. This problem is particularly acute in the case of necessary beings that are mutually exclusive, such as an omnipotent, omni-benevolent god and an omnipotent, omni-malevolent devil, or numbers and number-excluders.

A better solution is to be found in the rejection of (N2). It is (N2) that allows contradictions to be derived from apparent possibilities. But (N2) is supposed to be a conceptual truth. If (N2) is not true of platonic numbers, then there would seem to be no possibility of there being platonic numbers. It should not be so easy for anti-platonists to dispense with numbers. The

solution is not so much in the outright rejection of (N2), but in its modification, or perhaps in a more careful reading of it. We can retain (N2) so
long as the necessity in the consequent is a more restricted necessity than
logical necessity. This is similar to Kant's conclusion that there can be no
'absolute necessity of things', only 'conditioned necessity' (1781/1929, pp.
501-02). In the case of God, it has been suggested that God's necessity is
temporal necessity, since if God exists, then God must always have existed
and always will. God cannot be created or destroyed. Any world in which
God exists is a world in which God exists eternally.

Temporal necessity would seem to be part of the concept of platonic
numbers. There also seems to be more to the necessity of numbers than that.
The relations numbers bear to other numbers are necessarily unchanging. If
numbers exist, then seven must be the square root of forty-nine. Of course,
some properties and relations of numbers can change. Seven may have the
property of being Jane Campion's lucky number, but that can change. Nine
has the property of being the number of planets in the solar system, but if
Pluto should leave its orbit and head off into outer space, this would no
longer be the case. Platonists do not deny that numbers can change in this
(perhaps 'Cambridge') sense.

The precise nature of the necessity of numbers need not be fully
explicated here. Let us call it 'mathematical necessity' and symbolise it by
'$\Box_M$', reserving '$\Box$' for logical necessity. (N2) now becomes

(N2′) $\Box\,(\,(\exists x)Nx \supset \Box_M\,(\exists x)Nx)$

and the argument for (N4) and (N5) is no longer valid. Contradictions are
avoided and yet the notion that numbers exist necessarily is preserved, in a
sense. But now it is a contingent matter whether numbers exist with mathematical necessity in the actual world. The claim that this could be decided by
purely *a priori* methods becomes dubious. Discovering whether or not this
world is favoured by the existence of entities that might not exist would
seem to be a matter of observation. But how could we observe such entities
if they lack causal powers? The burden of proof is still on platonists to
explain why the existential causal condition (CE) should not apply in the
case of platonic objects.

In a recent paper, Hale outlines an argument for the necessary existence
of numbers (1994, p. 324). His argument relies on the assumption that
Frege's criterion of identity for cardinal numbers is necessarily true. Frege's
criterion is that for any F and G, the number of F's is identical with the

number of G's iff the F's and G's are in one-one correspondence. But since this criterion has existential import, to assume that it is necessarily true is to beg the question in favour of the necessary existence of numbers. From the necessary truth that F's are in one-one correspondence with themselves and Frege's criterion, we can deduce that something exists which is identical to the number of F's.

Anti-platonists need not deny Frege's criterion, but they need only accept that *if* numbers exist, then Frege's criterion is true of them. This conditional may be necessarily true, but it lacks existential import and we cannot deduce the necessary existence of numbers from it.[9]

7.10. COUNTEREXAMPLES TO THE EXISTENTIAL CAUSAL CONDITION

One way platonists could ease the burden of proof would be to offer counter-examples to (CE) in the case of non-platonic objects.

(CE) is very similar to a version of the causal theory of knowledge discussed by Mark Steiner:

(S4)   One cannot know anything about F's unless this knowledge (belief) is caused by at least one event in which at least one F participates. (1975, p. 115)

Steiner considers (S4) to be implausible, citing an example of scientists learning about long extinct animals by seeing their footprints in the forest floor (p. 116). According to his analysis, the animals participated in an event which caused the footprints, and the footprints participated in an event which caused the scientists to believe in the animals' existence; but the event in which the animals participated did not cause the belief. What happened, he maintains, is that an event 'led' to a condition and that condition was in turn 'responsible' for another event. But a condition, not being an event, cannot be a cause.

Call it what you will (event or condition), 'the soil's being depressed from the time the animals walked on it until the scientists saw it' seems to me to constitute a straightforward cause and effect. If the animals had been platonic, then there would have been no footprints for the scientists to observe.

---

[9] For a detailed criticism of the use of Frege's criterion to prove the existence of numbers see Field (1989, ch.5). See also my Section 10.2.

Suppose Ngaio puts poison in Gerald's coffee while he is not in the room (an event). The coffee is now poisonous (a condition). Ngaio leaves the room and some time later Gerald enters, drinks the coffee and dies (more events). Does anyone doubt that Ngaio's putting poison in the coffee caused Gerald's death? Could Ngaio's lawyer get her off the murder charge by calling Steiner as an expert witness?

I suggest that something like the following principle is sound:

(E)   If event E_1 causes condition C, and C is instrumental in event E_2's causing event E_3, then E_1 is a (partial) cause of E_3.

In other words, even if conditions are not themselves causes, they can make legitimate connections in a causal chain. Steiner's extinct animals are red herrings.

I turn now to a counterexample that will require an adjustment to (CE). We can know that at least one tallest spy exists (on earth), even though we may be causally isolated from every event involving a tallest spy.[10] We can infer this knowledge from knowing that at least one spy exists and that the earth is finite. Similarly, we can know that heavier-than-average screwdrivers exist without any causal contact with such screwdrivers. Appropriate causal contact with any two screwdrivers of different weights will suffice. A tallest spy must be a spy, a heavier-than-average screwdriver must be a screwdriver, and so on. So the causal constraint need only require that in order to know that a tallest spy or heavier-than-average screwdriver exists, one must have causal contact with an event in which at least one spy or screwdriver participates. Reformulating (CE) to accommodate this weaker requirement in an elegant way is a little tricky. I outline two options. Neither is really necessary, as the paradigmatic examples I have discussed should be sufficient to indicate what is required for the constraint to be met.

Let us call objects such as tallest spies or heavier-than-average screwdrivers *comparative* objects. In each case, an object qualifies as a comparative object in virtue of a comparative relation it has with the members of a set of which it is a member. A tallest spy must bear the taller-than (or equal-to) relation to all spies. Call the members of this larger set the *base* objects with respect to the comparative object. It may not always be clear what the base objects are for some comparative objects. Is the greenest valley chosen from all valleys or all green valleys? There could be a tallest spy even if

---

[10] The somewhat inelegant expression 'a tallest spy' is necessary to cover the possibility of a dead heat for the role of tallest spy.

there were no tall spies, but could there be a greenest valley if there are no green valleys? It depends on the speaker's intended meaning. We shall have to rely on context to discover what the speaker intends.

If we define the base objects for non-comparative objects as those objects themselves, then we can reformulate (CE) as:

(CE′)   We cannot know F's exist unless our belief in their existence is caused by at least one event in which a base object with respect to F's participates.

A further curlicue is necessary to allow for knowledge of base objects which are themselves comparative objects, for example, the shortest, heavier-than-average screwdriver, but I shall omit this.

Alternatively, we can reformulate (CE) as:

(CE′′)   If F's are non-comparative objects, then we cannot know F's exist unless our belief in their existence is caused by at least one event in which an F participates.

We need only add the extra premise: 'Platonic objects are non-comparative objects' for my original argument to be valid:

(CE′′)   If F's are non-comparative objects, then we cannot know F's exist unless our belief in their existence is caused by at least one event in which an F participates,

(C2)   Platonic objects cannot participate in events,

(C+)   Platonic objects are non-comparative objects,

therefore,

(C3)   We cannot know that platonic objects exist.

Other counterexamples may arise from situations like the following. Suppose we have a well-confirmed law that tells us that if a certain kind of atom A were sufficiently irradiated with neutrons it would produce short-lived radioactive particles of kind B. We might, at some time, have no reason to believe that A-atoms have ever been sufficiently irradiated and so have no reason to believe that B-particles have ever existed. If we subsequently bring about or observe a massive irradiation of A, then could we not then know that B-particles were produced even though they decayed too rapidly for us to detect them directly?

A similar, but more mundane, example might be as follows. You plan to make some new objects, objects that have never existed before. If you make your intentions sufficiently clear and convincing to me, it may be possible that I come, not only to believe, but also to know that such objects will exist. Later, without any further causal contact with you or the new objects, perhaps I could know that such objects do exist.

A case can be made for rejecting these as examples of existential knowledge. Discoveries like those of germanium, the top quark and Neptune suggest that scientists would not be satisfied that B-particles had existed until they had detected them by way of a causal interaction with events in which those particles had participated. This would probably be done by searching for predicted effects of the particles, rather than interacting with them while they still existed. Similarly, we may doubt that I know that your new object actually exists until I have perceived it or some of its effects. In the case of a new object that is no more than a simple but novel recombination of known objects, existential knowledge without causal interaction may be possible, but (CE) could be modified to allow for such cases without giving comfort to platonists. Call the constituents of a simple combination (one in which the constituents remain relatively unmodified) the 'robust' constituents of that combination. The condition on existential knowledge of F's then requires that our belief be caused by at least one event in which an F participates, or by events in which each of the robust constituents of F's participate. Knowing, without causal interaction, of the existence of objects with previously unknown physical, chemical or nuclear constituents or properties would still be ruled out, and rightly so.

However, if it were to be insisted that cases like that of the B-particles could be possible cases of knowledge, then (CE) may plausibly be extended to allow for knowledge acquired via a common cause.[11] In the B-particle case, our belief in the existence of B-particles would be causally connected to events giving rise to B-particles via the common cause of the irradiation of A-atoms. The causal condition so modified would require that our belief in F's be caused by at least one event in which an F participates, or by at least one event that causes an F to exist.

It might be argued that this modified condition is trivially satisfied if something like the big bang caused the existence of the entire material uni-

---

[11] I have already noted (in Section 4.5) that Goldman (1967, p. 364) suggests such a modification for his global condition on knowledge.

verse. Any belief that we have is caused by an event that caused the existence of all material objects. Platonists could object that the condition only rules out knowledge of objects that lack causality, and that this smacks of question begging against platonism. The requirement on permissible common causes needs to be tightened. I suggest that knowledge via a common cause is only possible if (i) we have considerable knowledge of the event which constitutes the common cause and (ii) that event is a close proximate cause of the known object. Scepticism concerning the existence of B-particles would increase the further down the causal chain from the irradiation event they supposedly came into existence. Incorporating (ii) suffices to meet the objection of question begging.

The following revision of (CE) incorporates all of the suggested modifications, but is more baroque in form than in substance:

(CE*)   If F's are non-comparative objects, then we cannot know that F's exist unless our belief in their existence is caused by:
  (a)   an event in which an F participates, or
  (b)   events in which each of the robust constituents of F's participate, or
  (c)   an event that proximately causes an F to exist.

7.11. FURTHER OBJECTIONS

I have been arguing that we cannot know that platonic objects exist. It might be argued that platonism could be true and we could have platonic knowledge even though we may lack the knowledge that platonic objects exist. Someone could know that cats exist and know much else about cats without knowing that mammals exist, even though all cats are mammals. Similarly, someone could know that numbers exist without knowing that platonic objects exist. But if numbers are acausal objects, then the belief that they exist would not meet the causal constraint (CE*). If platonism is true, then at some level we must have knowledge of the existence of objects that are acausal, if we are to have platonic knowledge.

Another possible objection is that I have overlooked the *de dicto/de re* distinction. It is possible to have *de dicto* existential knowledge without having *de re* existential knowledge and vice versa. For example, if I hear sounds at night that I cannot identify but which are in fact made by frogs in the bottom of my garden, then I know (*de re*) of the frogs in my garden that they exist, but I do not know (*de dicto*) that frogs exist in my garden. Alternatively, if a reliable witness informs me, I may know (*de dicto*) that there are frogs in my garden, but I may not know (*de re*) of any of those particular frogs that they exist.

Could it be that the causal requirement applies to only one kind of knowledge and that platonic knowledge is of the other sort? In fact, the causal requirement applies to both sorts of knowledge and blocks knowledge of platonic objects in both cases.

Suppose all platonic knowledge is *de re*. For example, we can know of the number 5 that it is prime, but we cannot know that 5 is a prime number. This is, of course, preposterous. If we do have platonic knowledge, then most, if not all of it, must be *de dicto*. In any event, the unmodified (CE) blocks all *de re* knowledge of acausal objects. *De re* beliefs arise from acquaintance and acquaintance is a causal process. Someone cannot be said to be acquainted with an object unless contact with the object has caused some change in the subject.

Suppose all platonic knowledge is *de* dicto. I have assumed that all the existence claims discussed have been *de dicto*. I noted that it is possible to know (*de dicto*) that the tallest spy exists without having causal contact with that object. However, it turns out that this is only possible because the knower has *de re* knowledge as well as *de dicto* knowledge of a class of objects of which the tallest spy is a member. In brief, my argument has been that all existential knowledge involves some *de re* knowledge, and platonic knowledge is not possible without *de re* knowledge of platonic objects.

7.12. DEFENDING THE ELEATIC PRINCIPLE

Mark Colyvan (1998a) investigates and criticises what he calls the Eleatic Principle. This is the principle that we are only justified in believing in the existence of those entities to which causal powers can be attributed. He seeks to undermine the principle by examining the motivations and arguments that have been, or might be, employed in its favour and demonstrating that those arguments do not provide adequate justification for such a principle. Insofar as (CE*) may be seen as a version of the Eleatic Principle, it is instructive to see whether any of his arguments undermine my arguments for (CE*).

Colyvan's aim is to criticise 'the motivation for any formulation of the principle' (p. 324), so at least some of his criticisms should undermine my arguments if he is to achieve that aim. Indeed, some of his criticisms are specifically aimed at my particular formulation, but those are by no means the only criticisms that seem germane.

He starts by specifying a particular version of the principle, but only by way of providing a useful example. Unfortunately this particular formulation immediately introduces some confusion. His version is:

(EP)   An entity is to be counted as real if and only if it is capable of participating in causal processes. (p. 324)

For an entity to be 'counted as real' apparently means simply that we are justified in believing in its existence. The idea that an entity's being capable of participating in causal processes is *sufficient* to justify our belief in its existence is clearly false. There are presumably many as yet unthought of, let alone undiscovered, yet causally potent entities in existence. We are not (at present) justified in believing in their existence and some of them we (and our ancestors) will never be justified in so believing. Colyvan recognises that there is a problem here and turns to the genuine issue of whether causal capability is a *necessary* condition for justified existential belief. So his exemplar of the Eleatic Principle should perhaps read:

(EP*)   We are justified in believing that an entity exists only if it is capable of participating in causal processes.

A further problem arises from the formulation of the Eleatic Principle as a criterion for justified belief. Epistemic justification is always relative to epistemic circumstances. Unless those circumstances are specified in some way, then it is unclear what precisely is being claimed by (EP*). If it refers to any possible epistemic justification, then it is false. If testimony can be a source of justification, it is possible for someone to be justified in believing, on the basis of expert testimony, that platonic entities exist. Colyvan does not believe that (EP*) can be so easily dispatched, so presumably he has some more specific epistemic circumstances in mind. Or maybe he has a notion of epistemic justification that is circumstance-independent. In any event, there is at least a danger of equivocation when justification is appealed to in this way. I say more about the relationship between justification, knowledge and existence later in this section.

The first argument Colyvan examines is an inductive argument to the effect that all the entities that we accept uncontroversially as real (that is, believe to exist) are capable of causal activity, while those that we reject are causally impotent. He dismisses the argument on the grounds that the uncontroversially real entities also share other properties, such as spatio-temporal location or having a positive rest-mass. But that point by itself is not enough to undermine the inductive argument. It might be that these other properties are also necessary to justify existential beliefs. It turns out, though, that Colyvan is actually suggesting that if these other properties demarcate the controversial from the non-controversial then a different set of entities will be counted as real. Spatio-temporal location will allow in space-time points while non-zero rest-mass will exclude photons. What he doesn't notice is that these demarcations do not separate the controversial from the non-controversial. The reality and/or nature of space-time points is controversial while that of photons is not. The reason for this difference has more than a little to do with their respective causal capacities. The existence of photons is relatively uncontroversial, given our detection of them by causal interaction. Debate over the nature and existence of space-time points gives rise to the competing positions of relationalism (which denies their existence) and substantivalism (which asserts their existence, but divides over whether or not they have any causal capacity).[12]

Colyvan's point does draw our attention to the issue of why advocates of the Eleatic Principle choose to focus on causality rather than any other property. He suggests the obvious answer. '[I]t is by virtue of an entity's causal efficacy that we have epistemic access to it.' (Colyvan 1998a p. 316) This leads to his discussion of the second kind of argument (the epistemic argument) as a motivation for the Eleatic Principle.

First he suggests that the epistemic argument has to be seen as motivating a requirement for causally activity *with humans*. This is surely right, as it is the epistemic access of human knowers that is at issue. He points to the existence of stars and planets outside our light cone as counterexamples to this requirement. But these do not undermine my (CE*) since our belief in their existence is caused by an event (the Big Bang) which was their proximal cause. In this case, the proximity does not need to be especially close, because it part of our understanding of the nature of the Big Bang that it must have been the cause of stars and planets outside our light cone.

Colyvan then turns to my argument as I present it in Section 7.7 and in Cheyne (1998). He sees my formulation as a move to one in terms of *kinds* of entities and sees a difficulty with this move (Colyvan 1998a fn.9) but I cannot see that Colyvan's own formulation (EP) can be understood *except* in terms of kinds of entities. Any question as to whether or not an entity exists is a question as to whether or not an entity meeting a certain description

---

[12] See Field (1989, Ch. 6).

exists. In other words, whether or not a certain kind of entity exists. And that is so, whether or not the description can be met by only one entity or by many. The difficulty he raises is that of deciding whether (in 1846) Neptune was a new kind of entity. After all, it was already accepted that planets exist. But what was at issue was the existence of an eighth planet or of a planet with an orbit beyond the orbit of Uranus. My formulation may not explicitly allow for singular cases, but can be readily adjusted to do so.

Colyvan takes issue with my Neptune example because 'we certainly had causal contact with Neptune prior to Galle's visual contact' (p. 318), since we had indirect contact via its disturbance on the orbit of Uranus. He claims that the moral to be drawn from such examples is 'don't settle for indirect evidence if you can do better.' (p. 318). Maybe so. But this in no way undermines the claim that issues of existence are to be settled by making the right sort of causal contact with the postulated entity. My argument is that the appeal to such contact supports the claim that causal contact is a necessary condition for existential knowledge. Colyvan's non-contentious observation does nothing to undermine my argument in support of that claim.

Colyvan's comments on my Mendeleeff example (Section 7.7) are of more interest. I claim that any belief in the existence of germanium prior to 1887, although true, would not be knowledge, but could only be a 'lucky guess'. Colyvan objects that even if we would not count such a belief as knowledge, 'it seems extremely harsh to call [it] a "lucky guess"' (p. 319). He points out that Mendeleeff could have justified his belief by appealing to the past predictive success of his periodic table or by some sort of appeal to symmetry. So Mendeleeff could have had a justified belief in a substance with which he had had no causal contact, and that (according to Colyvan) constitutes a counterexample to the Eleatic Principle as formulated by (EP*).

First, I note that germanium is not an example of an entity incapable of participating in causal processes. But let that pass. What Colyvan's observations highlight is the unsatisfactoriness of formulating such an epistemic principle in terms of justified belief. As I noted above, whether or not a belief is justified depends on the epistemic circumstances of the believer. Mendeleeff could be justified in believing well nigh anything, given an appropriate epistemic state. He could have been justified in (falsely) believing that there are more than 120 different, naturally occurring elements on Earth or in (truly) believing that the 17th President of America would be assassinated.

Suppose Mendeleeff was inclined to believe (prior to 1887) that germanium exists. I can concede that calling his belief a 'lucky guess' is a bit harsh, without conceding that this constitutes a possible counterexample to

my causal condition on existential knowledge. Knowledge is more than justified true belief. (See my Chapter 4.) That Colyvan agrees with this is implicit in his comments. Mendeleeff's belief, even if justified, would still have a
fortuitous element sufficient to deprive it of the status of knowledge.
Furthermore, if Mendeleeff also believed, as seems likely, that the right sort
of causal contact with germanium would be necessary to establish its
existence, then this belief would undermine whatever justification he may
have had for believing in the existence of germanium.

Colyvan concludes his discussion of the epistemic argument for the
Eleatic Principle with a curious footnote, 'In fairness to Cheyne, though, he
is interested in a causal criterion of *existence*, whereas I [Colyvan] am
interested in a causal criterion of *justified belief*.' (fn. 12). As I have just
made clear, I am interested in a causal criterion of *existential knowledge*. A
causal criterion of *justified belief* cannot be defended, while a causal criterion of *existence* is nonsensical. For an entity to exist it need only, well,
exist or be. I do not deny that platonic objects could exist, simply that we
cannot know of their existence and that when we realise this we are no
longer justified in believing in their existence.

The next argument in support of the Eleatic Principle that Colyvan
examines is the argument from causal explanation. This argument is to the
effect that we ought not to postulate the existence of causally inert entities
because the postulation of such entities can have no explanatory value.
Colyvan confronts this argument directly by presenting two explanations in
which, he claims, causally idle entities play an important role. Unfortunately,
he does not take the trouble to make it clear why he supposes that the
success of these explanations depends on postulating the *existence* of such
entities.

The first example concerns antipodal weather patterns. How can we
explain the fact that at some time t_0 there are two antipodal points on the
earth's surface with exactly the same temperature and atmospheric pressure?
We could explain this coincidence by tracing the causal history of the
weather patterns prior to t_0, but a better explanation would appeal to the
proof of the topological theorem that states that there are always two such
antipodal points. According to Colyvan, 'this explanation makes explicit
appeal to non-causal entities such as continuous functions and spheres.' (p.
322)

But what the proof demonstrates is that if a spherical object has two
quantities that change continuously across its surface, then at all times there
will be at least two antipodal points at which the values of the two quantities
coincide. If we assume that the earth and its weather sufficiently approx-

imate such an object, then the explanation follows. The existence of caus-ally-inert objects need not be postulated. The topological theorem and its proof do not assume the existence of anything. When we apply the topo-logical theorem to a particular case, we appeal to the existence of a causally active object. *Prima facie*, the existence of a spherical planet does not require the existence of an acausal sphere, and temperature and pressure can vary continuously over the surface of that planet without the existence of a mathematical function. If there is some sort of necessary co-existence here, then an argument to that effect is required.

Colyvan's second example is an explanation of Fitzgerald-Lorentz contraction, that is, the reduction, according to special relativity theory, of the length of a body in the direction of its motion. The explanation is that, rather than actually contracting, the body is simply rotating and translating in Minkowski space-time. He claims that it is 'obvious...that this is a purely geometric explanation featuring such non-causal entities as the Minkowski metric and geometric properties of Minkowski space' (p. 323). Because it is preferable to leave it open as to whether or not properties need be included in our ontology, this might be better expressed as the claim that 'this is a purely geometric explanation featuring an acausal entity, namely, Minkow-ski space'. But this claim is not at all obvious. What I take to be the clear message here (a message that is reinforced by general relativity theory) is that space-time and subsets of space-time are not causally inert. Bodies will move in certain ways because they are moving in Minkowski space rather than moving in a different space with different causal properties.

Colyvan fears that attributing causal powers in this way will seriously blur the distinction between the causal and the acausal, emptying the term 'causal', and hence the causal criterion, of any interesting significance. He seeks to reinforce this point with his discussion of explanations as to why a square peg of side length l will not fit into a round hole of diameter l (p. 324). Two explanations are offered. There is an apparently non-causal explanation, which appeals to the geometry of the situation, that is, to the squareness of the peg and the roundness of the hole, and a causal explan-ation, which appeals to the resistance offered by the overlapping portion of the peg. According to Colyvan, the first explanation is more informative than the second, so cannot simply discard it in favour of the second. He sug-gests that the way out for causalists is to accept squareness and roundness into their ontology as causal properties on the grounds that, although not causally efficacious, they are causally relevant in such cases. But this would so attenuate the notion of 'cause' that the causal criterion would be pre-served in name only.

This example can be treated in a similar way to the weather-pattern example. We have two separate explanations, with two different explananda. That they may appear to have the same explanandum arises from a certain ambiguity of the word 'fit'. The first explanation is the geometric proof that any square object of side length l will overlap any round object of diameter l. No existential claims need be made and the result applies to any objects, whatever their causal status. Here, 'fitting' is a matter of 'not overlapping'. The second explanation applies only to overlapping objects with certain causal powers and explains why one will not pass into or through the other. Here, 'fitting' involves 'not blocking the passage of'. Once again, the overall explanation does not appear to depend on invoking the existence of causally inert entities, whether squares and circles, or squareness and roundness.

Colyvan's discussion of the above examples was intended to counter the claim that only causally active entities can have explanatory power. He hopes to achieve this by presenting successful explanations in which the existence of causally inert entities is invoked. My response was to deny that the existence of such entities is invoked. Either the entities are causally potent or their existence is independent of the success of the explanation. Talk of acausal entities may have a useful role in successful explanations. It is just that such talk need not (and should not) carry with it any serious ontological commitment. Other examples of useful talk of acausal entities involve idealisations. Frictionless planes, non-turbulent laminar flow, perfect circles, and the like are clearly fictional, but no one can doubt their explanatory power (as opposed to their causal power). Colyvan concedes this, but points to our talk of numbers, sets, vectors, etc which do not appear to be playing the same role as idealisations in our explanations. Having attempted to repudiate a number of arguments and motivations in support of the Eleatic Principle, it emerges that his main reason for rejecting such a principle is the Quine-Putnam indispensability argument. In this section, I have argued that his repudiation (in particular, of my CE*) does not succeed. I respond to the Quine-Putnam indispensability argument in Chapter 11.

A final point. Colyvan supposes that some supporters of the Eleatic Principle seek to restrict the principle of inference to the best explanation. I find this somewhat perverse. The causalists' position is not that we should accept the ontological commitments of our best explanations except when doing so would lead us to platonism. Rather, they claim that (all things considered) an explanation that invokes the existence of platonic objects will not be an explanation that we should accept. Interestingly, Colyvan comes close to accepting this latter principle, but only as a '"rule of thumb" that may not be the final arbiter in such matters' (p. 332). Nor *may* it (we are all fallible epistemic agents), but it is the principle that we should accept in the meantime.

7.13. NON-NATURAL SCIENCE AND THE CAUSAL CONDITION

JC Beall (2001) offers further criticisms of (CE) and my arguments for it. Recall:

(CE)   We cannot know that F's exist unless our belief in their existence is caused by at least one event in which an F participates,

Beall claims that two claims need to be distinguished:

($CE_{natural}$)   CE is true of entities postulated within natural science.

($CE_{non-natural}$)   CE is true of entities postulated within non-natural science.

By non-natural science, he means mathematics, logic, semantics, and the like. Beall grants ($CE_{natural}$) but argues that, not only do we have no reason to accept ($CE_{non-natural}$), we have reason to reject it. Indeed, he argues that my examples from the history of science provide such a reason. Common to the examples I cite, according to Beall, is the natural scientists' expectation that causal interaction should occur with their posited entities. This is because these entities are initially posited to play a causal role in order to fill a causal explanatory gap. On the other hand, the posits of non-natural scientists are a different matter. Mathematicians postulating the existence of imaginary numbers or semanticists postulating the existence of propositions do not expect to have causal interaction with these entities because they are not posited as having causal powers.[13] This lack of expectation Beall takes to be a reason for rejecting an existential causal condition in mathematics and other non-natural science.

Beall's argument misses the point. The key feature in my examples is not, as he supposes, the scientists' *expectation* of causal interaction. An expectation of causal interaction might play a role in an argument for a sufficient condition, but it cannot play such a role in an argument for a necessary con-

---

[13] Beall incautiously claims that these non-natural scientists are, in fact, postulating the existence of acausal entities, but this is to take a large step in the direction of begging the question. For the purpose of his argument, all he need claim is that causal powers are no part of their postulation.

dition. Consider the following simple example. Suppose I claim that there are blackbirds in my garden. I may expect to hear them singing, and my hearing them sing (thus, meeting my expectation) may play a role in confirming their existence, but it is not necessary that I hear them sing in order to confirm their existence. The key feature in my examples is the *requirement* of causal interaction if the existence of the scientists' posited entities is to be confirmed. It is this requirement that points to the necessity of the causal condition. Or, to put it another way, their implicit acceptance of a causal condition on existential knowledge explains the scientists' behaviour in seeking causal interaction.

Suppose causal interaction is not required for confirming the existence of certain (non-natural?) posited entities. Then we would have a puzzle. If we can know that such entities exist, then some condition must suffice to confirm the existence of these entities. So, why shouldn't the same condition suffice to confirm the existence of entities that happen to have causal powers? If causal interaction is not necessary for existential knowledge of entities that are not posited as having causal powers, why should it be necessary for entities that are posited as having causal powers?

Beall has not given us a reason for rejecting ($CE_{non-natural}$). Even if we remain agnostic towards ($CE_{non-natural}$), it doesn't follow that we can or do have existential knowledge of platonic objects, something Beall acknowledges later in his paper. There is still a burden of proof on platonists to show how such knowledge is possible. It is not even clear that we should accept the natural/non-natural distinction in the first place. To do so may be to accept implicitly that the so-called non-natural scientists are in the business of postulating the existence of platonic objects. Whether they are or not is part of the controversy in which we are engaged.

To sum up, in this chapter I have argued that in the case of non-controversial existential knowledge of objects, a causal constraint is necessary and that we have no grounds for supposing that it is not necessary in the case of platonic knowledge. We cannot know that objects exist unless they have causal powers.

CHAPTER 8

THE BURDEN OF PROOF

8.1. PLATONISM'S BURDEN OF PROOF

Suppose platonists could counter my causal objections and demonstrate that it is *possible* for us to have knowledge of acausal objects. This would still not establish the platonist position. First, I argue that the burden of proof is on platonists to explain how we *do* have such knowledge. Then I examine platonist attempts to provide such an explanation. I show that they are inadequate.

Paul Benacerraf (1973) argues that our best semantic theory for mathematical propositions is in conflict with our best epistemology. (See my Section 1.3.) The semantic theory he has in mind entails platonism and the epistemological theory is a causal theory of knowledge. Anti-platonists have adopted his argument in order to challenge the viability of platonism. In outline, their argument is:

1. Platonic objects are acausal

2. The causal theory of knowledge

∴ 3. We cannot have platonic knowledge

∴ 4. Platonism is untenable.

Since Benacerraf's paper, strong causal theories of knowledge have fallen from favour, even among anti-platonists. Hartry Field declares that 'almost no one believes [them] anymore' (1989, p. 25). Alvin Goldman, the promulgator of the best known of such theories, no longer subscribes to it (1976, p. 771).

Platonists have been quick to point this out. (For example, Burgess 1990, p. 6.) But the causality objection to platonic knowledge requires only a causal constraint on knowledge, and the source of such a constraint need not be a 'full-blown' causal theory of knowledge. Although I have appealed to some arguments that support a causal theory of knowledge, I have also presented independent reasons for accepting that there is a causal constraint on knowledge. For that matter, although Benacerraf does mention that he favours 'a causal account of knowledge', his subsequent remarks are best seen as an argument for a causal constraint (1973, pp. 412-14).

In outline, my anti-platonist argument is:

> A1.   Platonic objects are acausal

> A2.   The strong causal constraint on knowledge (SC*) or the causal constraint on existential knowledge (CE*)

∴   A3.   We cannot have platonic knowledge

∴   A4.   Platonism is untenable.

An obvious platonist response to such an argument is to use it as a *reductio* of the causal constraints. Platonists argue that since we undoubtedly have mathematical knowledge, it is the causal constraints on knowledge that are untenable. 'Our knowledge of mathematics is ever so much more secure than our knowledge of the epistemology that seeks to cast doubt on mathematics.' (Lewis 1986, p. 109) But this counter-argument requires a further premise if it is to be valid.

> P1.   We do have mathematical knowledge

> P2.   Mathematics entails the existence of acausal objects

∴   P3.   The causal constraints on knowledge, (SC*) and (CE*), are untenable.

Now we can see that it is the relative plausibility of A2 and P2 that is in dispute. Anti-platonists need not be committed to the difficult task of casting doubt on the security of our mathematical knowledge. Even Field, who accepts P2 insofar as it relates to standard, classical mathematics, accepts that we do have some mathematical knowledge, albeit that it is a certain kind of logical knowledge which does not entail the existence of platonic objects (1989, ch. 3). I have argued extensively for A2. A case can be made for P2. But it is not as obvious as platonists would like us to believe. It is a highly theoretical claim. It is rarely encountered in the writings of mathematicians, pure or applied, from the earliest to the most recent. Many ontological claims in the sciences are quite 'up-front'. Scientists take pains to describe what sort of entities barnacles, electrons, comets, acids, and so on are. When there is doubt as to the nature of certain entities (for example, are photons wave-like or particle-like?), it becomes an issue of serious discussion, debate, and investigation. Not so with talk of entities which may, or may not, be abstract. Here I refer not just to talk of mathematical entities, but to talk of such things as properties, functions, and species. Such talk is, on the whole, ontologically unreflective. Mathematicians who have not had such philosophical issues drawn to their attention are often surprised by the claim that their statements entail the existence of acausal objects. After all, their professional lives are devoted to discovering what is entailed by mathematical claims. Here is an entailment that does not seem to be of serious interest to them. It may just be that it is not intended by what they say. Perhaps it arises from a misinterpretation of the meaning of mathematical language.

This issue arises with respect to the indispensability thesis and is discussed in more detail in Section 11.3. There is another problem for P2. Together with an uncontroversial assumption, P1 and P2 imply that there is a process that is a source of platonic knowledge, in particular, a source of the knowledge that platonic objects exist. The uncontroversial assumption is that any knowledge arises as the result of some process. Here, 'process' may be understood in a broad sense. Even on a simple coherentist account of knowledge, the acquiring of a coherent set of beliefs by whatever means may count as a process. I argued in Chapter 3 that any plausible account of knowledge must include the requirement that the beliefs concerned have an appropriate etiological history, but that requirement is not intended here. However, we have lost our grip on the notion of knowledge if we suppose that it can arise from nowhere. We would also have an inadequate notion of knowledge if we supposed that the processes giving rise to individual items of knowledge had nothing in common. In other words, the assumption is to be read as claiming that each item of knowledge arises as the result of some type of process. Not the same type for all knowledge, of course, nor even for all cases of platonic knowledge. But anyone committed to P1 and P2 should also be committed to the claim that there is at least one type of process that gives rise to platonic knowledge.

If mathematical knowledge is as secure and ubiquitous as mathematical platonists claim, then it is reasonable to suppose that evidence is available as to what sort of process or processes account for platonic knowledge. A total absence of such evidence would cast serious doubt on the existence of such processes and hence on platonic knowledge itself. That is why I say that there is a burden of proof on platonists to explain how we do have platonic knowledge.

Hartry Field makes a similar point, but makes it in such a way as to minimise his commitment to any epistemological position, even one as minimal as that which I assume. His challenge for platonists is to explain the *reliability* of the mathematical beliefs of experts. Platonists must:

> provide an account of the mechanisms that explain how our beliefs about these remote entities can so reflect the facts about them. The idea is that *if it appears in principle impossible to explain this*, then that tends to *undermine* the belief in mathematical entities, *despite* whatever reason we might have for believing in them. (1989, p. 26, his emphasis.)

Talk of the controversial notions of truth and facts can be avoided. What mathematical platonists must accept and explain is the claim that for most mathematical sentences that you substitute for 'p', the following holds:

(M)   If mathematicians accept 'p' then p. (p. 230)

He could even avoid talk of the mentalistic notions of belief and acceptance by revising (1) to:

(M′)   If mathematicians assert 'p' then p.

Platonists must claim that the world has a certain feature, whether that feature is to be cashed out in terms of knowledge, reliable beliefs, or reliable assertions, and that claim carries with it the obligation of demonstrating that an explanation of the feature is at least possible.

8.2. RELIABILISM AND PLATONISM

Before turning to platonist attempts to meet this obligation, I discuss another anti-platonist argument. Even though this argument is ultimately unsuccessful, it does throw further light on the platonists' burden of proof.

A current, fashionable epistemology is reliabilism. From an anti-platonist point of view, it would be nice to be able to argue as follows:

1.   Platonic objects are acausal

2.   The reliabilist theory of knowledge

∴   3.   We cannot have platonic knowledge

∴   4.   Platonism is untenable.

Unfortunately, reliabilism does not overtly require a causal connection between the objects of knowledge and the knower's belief (see Section 4.9), so *prima facie*, it is not incompatible with platonism. Albert Casullo (1992) examines reliabilist accounts of knowledge and argues that if such accounts are to be strong enough to be plausible, then they will not allow knowledge of platonic entities. Casullo identifies two kinds of reliabilist accounts, reliable indicator and reliable process. I discussed Armstrong's version of the reliable indicator account in Sections 6.4 and 6.5. My conclusions are essentially the same as Casullo's. Such an account is not compatible with our having platonic knowledge.

A basic account of process reliabilism is:

(RP)   S knows that *p* iff
  (a)   *p* is true,
  (b)   S's belief that *p* is caused by a reliable process.[1]

Unlike the reliable indicator account, the causal theory and the other accounts of knowledge I discussed in Chapter 6, process reliabilism does not require a particular kind of relationship between S's belief that *p* and the fact that *p*. So (RP) appears to be congenial to platonism, as it does not explicitly require that the objects of belief play any role (causal or otherwise) in the generation of the belief.

Casullo argues that (RP) is too weak, as it does not require that S's belief be justified. Suppose S has a belief that satisfies (a) and (b) but one or more of the following conditions obtains:

(1)   S has good reason from another source to believe that *p* is false,
(2)   S has good reason to believe that the process that she believes caused her belief is unreliable,
(3)   S has good reason to believe that there is no such process as the one she believes caused her belief,

---

[1] Here, and in the following, I have paraphrased Casullo's versions of the conditions for knowledge.

then S's belief will not qualify as knowledge.

Casullo suggests (from Alvin Goldman 1986, p. 63) that (RP) also requires:

(c)   S's belief that p is not undermined by S's cognitive state.

He then considers the possibility of beliefs acquired by clairvoyance. Suppose clairvoyance under certain conditions is a highly reliable process but, as is actually the case, any evidence for the existence and reliability of it is weak and controversial. Someone with clairvoyant abilities, but no relevant beliefs about the process of clairvoyance, could meet RP conditions (a), (b) and (c) but, according to Casullo, would not have knowledge. He suggests further conditions to strengthen (RP):

(i)   S's belief should not be undermined by what S would (or could) be justified in believing given his cognitive state.

(ii)   If S's belief is produced by process R, then S must be justified in believing that his belief is produced by R.

(iii)   S's belief should not be undermined by other evidence available within S's epistemic community.

Casullo (1992, p. 583) argues that, even if clairvoyance is a reliable process, it cannot yield knowledge (at present) because any belief produced by clairvoyance will violate at least one of the above conditions. And, given our current knowledge of reliable and unreliable belief-forming processes, the requirements of his strengthened version of process reliabilism together with the causal inertness of abstract objects lead to the conclusion that 'mathematical intuition' does not provide us with knowledge of such entities.

Casullo's justification requirements for knowledge are too strong. I agree that if S has actual or potential beliefs that undermine her belief that p (cases such as (1), (2) and (3) above), then S may not know that p, but S could have knowledge via a reliable but mysterious process, even though S's epistemic community has evidence that there is no such process. This is particularly the case if S is epistemologically naïve or apathetic, so that she lacks any substantive beliefs about belief-forming processes. The little boy knew that the emperor had no clothes on, even though he was ignorant of his epistemic community's belief that the emperor was wearing clothes cut from a cloth so fine that only the wise could see it. If the conditions are right, the epistemologically naïve can acquire knowledge that may be denied to the epistemologically sophisticated under similar conditions.[2] Some may find this dubious. So it would be if it implied that epistemological naïvety is a good epistemic strategy. But it does not imply this. It is quite consistent with the view that, in general, epistemological sophistication will enhance our ability to acquire knowledge, making it more likely that we will discover truth and avoid falsehood.

It may be that (RP), even with condition (c), is too weak, but Casullo's further conditions make it too strong, so we still lack an argument that a plausible reliable-process account of knowledge disallows platonic knowledge. However, Casullo's argument does lend weight to the 'burden-of-proof' argument against the platonists. Although reliabilism does not rule out the possibility of knowledge of platonic objects, it does require that there *be* a reliable process if we actually have such knowledge. If we are to be platonists, then we must be justified in believing that such a process exists.

One further problem with Casullo's arguments is that they apply only to non-inferential knowledge (pp. 562 & 571). He doesn't explain this restriction. Perhaps he believes that if you cannot have non-inferential knowledge of platonic objects, then you cannot have inferential knowledge of them.[3] This is certainly so for inferential knowledge gained by *deduction*. We can only deduce propositions with existential import from axioms that have existential import. But consider the kind of inference involved in the process which Quine and Putnam claim yields platonic knowledge (to be discussed in Chapter 11). Many would consider this to be a non-deductive form of inference. Quine and Putnam argue that it simply employs the inferential methods of empirical science. If so, it appears to meet Casullo's very strong justification requirements. Or it may be that inferential knowledge does not need to meet such strong conditions. Surely, correct inference can yield knowledge irrespective of one's epistemic community's theories of logic. We need an argument to the effect that the Quine/Putnam-style inference is not a reliable process, or at least that it has not been established, or we are not justified in believing, that it could be a reliable process for acquiring beliefs about platonic objects.

In the following chapters I examine four prominent accounts of how we come to have knowledge of the existence of platonic entities. These are, first, that we know of them by intuition; second, that our knowledge of them is 'conceptual'; third, that we know of them by a process of postulation and

---

[2] Carrier (1976, p. 250) and Lycan (1977, p. 122) make a similar point.
[3] Maddy (1990, p. 45) points out that anti-platonists often make this 'unspoken assumption'.

scientific confirmation; and fourth, that we can know of them by stipulation, if every possible platonic entity exists.

CHAPTER 9

PLATONIC KNOWLEDGE BY INTUITION

9.1. VARIETIES OF INTUITION

Intuition has been suggested as the means by which we gain platonic knowledge.[1] An immediate problem is that the term 'intuition' is employed with a wide range of meanings.[2] If these uses of the term have anything in common, it is that intuition is related to the acquisition of belief by a process that is apparently immediate and non-inferential. I shall identify those uses that are most often associated with the purported acquisition of platonic knowledge and show that none of them refers to a process that both exists and could perform as claimed. It is important to examine each notion separately. There is the danger that evidence for one intuitive process may be adduced as evidence for another, somewhat different, process. One process may genuinely yield some sort of knowledge, but be incapable of yielding *platonic* knowledge. The other, *if it existed*, might yield platonic knowledge, but any evidence for the former would, of course, be irrelevant to the existence of the latter.

Perhaps the commonest non-technical notion of intuition is that of intuition as apparently unjustified belief that seems immediate (not a result of inference) yet is accompanied by a feeling of conviction. We might call this 'intuition-as-hunch'. A detective may claim that she has solved a crime by intuition. Having considered a bewildering array of clues and suspects, she

---

[1] This chapter is based on Cheyne (1997b).
[2] Noddings & Shore (1984, ch. 1 & 2) give a useful historical survey of the varying use of the term in philosophy, theology, and psychology, although they do not make much progress in disentangling the various concepts. Bastick (1982, p. 25) identifies twenty properties associated with the general use of the term.

has an inexplicable hunch that the butler did it. Following up on the hunch, she eventually pieces together a convincing case against the guilty party. Similarly, mathematicians often claim that they discover mathematical truths by intuition, discoveries that are later confirmed by formal proof. We can have such hunches about anything and many of our hunches (perhaps a surprising number) turn out to be true. So this notion does not apply to any particular kind of belief-content and the existence of such intuitions cannot be denied.

Now it may be that no special faculty is at work here, at least, not one that yields knowledge. Successful hunches may be no more than lucky guesses. Perhaps detectives, mathematicians, and the rest of us tend to remember the successful hunches and to forget those that lead nowhere.[3] If so, successful hunches give no support to the claim that intuition is a source of platonic knowledge. Suppose, on the other hand, that there is such a faculty, and that it is a reliable source of true beliefs. Such a faculty could involve unconscious (and very rapid) inference. Or it could involve a direct apprehension of the state of affairs that comes to be known. So, the notion of intuition-as-hunch suggests two possible processes for knowledge by intuition.

A somewhat different process of intuition as direct apprehension is invoked by the claim that we have a special faculty for intuiting platonic entities or platonic states of affairs, rather than any state of affairs. Again, a more technical notion of intuition is invoked by the claim that intuition is part of the process of ordinary sensory perception. Finally, there is the claim that the truth of certain propositions may be intuited when we contemplate them.

We have five different processes that involve some notion of intuition and that are relevant to the claim that we can obtain platonic knowledge by intuition:

(1)   Intuition as unconscious inference (inferential intuition).
(2)   Intuition as direct apprehension of any state of affairs (ESP intuition).
(3)   Intuition as part of the process of ordinary sensory perception (perceptual intuition).
(4)   Intuition of the truth of certain propositions (cognitive intuition).

---

[3] 'And such is the way of all superstition, whether in astrology, dreams, omens, divine judgements, or the like; wherein men, having delight in such vanities, mark the events where they are fulfilled, but where they fail, though this happen oftener, neglect and pass them by.' Bacon (1620/1960, p. 51)

(5)   Intuition as direct apprehension of platonic entities or platonic states of affairs (direct platonic intuition).

These five processes appear to exhaust the possible ways for intuition to yield propositional knowledge of an external reality, and hence to yield platonic knowledge. I examine each in turn.

9.2. INFERENTIAL INTUITION

When mathematicians claim that they discover mathematical truths by intuition, they are usually talking about a sudden realisation accompanied by a firm conviction that a certain mathematical proposition is true. This mode of discovery contrasts with processes whereby mathematicians come to believe a mathematical proposition as the result of following a piece of reasoning or a proof (be it formal or informal). Discovery by conscious inference involves becoming convinced as the result of understanding why something is the case. Intuition, on the other hand, seems to involve conviction without such an understanding.

Suppose that this sort of intuition is no more than unconscious (and very rapid) inference. If so, then it must be the same process that allows me, when reading a detective novel, to have an intuition that the butler did it. The point is that 'intuition-as-unconscious-inference', if it works, works just as well in fictional and hypothetical realms as in any other. So we can agree that knowledge can be obtained by this process, without agreeing that it can yield existential knowledge. Such intuition allows us to 'see connections' that can, in principle, be made by conscious inference. It is simply achieving by unconscious inference that which can also be achieved by conscious inference. If the claim is that intuitive knowledge is no more than a special kind of inferential knowledge, then we need an account of how inference can yield platonic knowledge. It is inference, not intuition, which would be doing the important work in such an account. The question of whether inference alone can provide such knowledge is discussed in Chapter 11.

9.3. ESP INTUITION

Suppose that knowledge by intuition involves a direct apprehension of the state of affairs involved. This assumes some sort of clairvoyant or telepathic faculty. A platonist could argue as follows. There is evidence that we can directly intuit states of affairs by a kind of extra-sensory perception. For example, psychics foresee the future, perceive the location of distant objects, and read minds. There is no evident causal link with the states of affairs in such cases. Therefore, it must be a non-causal process. If we can make contact with concrete states of affairs by a non-causal process, then it should be possible for us to make contact with states of affairs that are causally inert by means of the same, or a similar, process.

There is little or no good evidence for such a process or faculty. An epistemology that relies on the results of paranormal research is on very shaky ground.[4] Besides, any evidence we have for the existence of telepathy or clairvoyance should most plausibly be regarded as evidence for a causal process the details of which are as yet undiscovered. How else are we to explain the matching of concrete (non-platonic) states of affairs with true beliefs about those states, unless we suppose that that it is the states of affairs that give rise to the beliefs? The evidence for ESP is meagre enough, without its being expected to give support for a mysterious and unknown non-causal process. The claim that platonic knowledge can be acquired by a non-causal process is not supported by the fact that we have successful hunches about a wide range of non-platonic states of affairs.

9.4. PERCEPTUAL INTUITION

The third notion of intuition to be considered sees it as part of the ordinary process of sensory perception. One argument for the involvement in perception of a process worthy of the title 'intuition' goes as follows. When I see that there is a tree in front of me or see that the tree in front of me is an oak, then I acquire beliefs that are immediate upon seeing the tree and do not seem to arise from any rational or inferential process. But seeing a tree is one thing, seeing that it is a tree or that it is an oak is another. The former is a case of perception-*of*, the latter are cases of perception-*that*. Perceptual intuition supposedly closes the gap between perception-of and perception-that.

Perceptual intuition, whatever it may be, seems to be a causal process. Light reflected from the tree enters the eye, causing activity in the visual cortex. That activity causes certain sensations, which in turn (no doubt mediated by other occurrent mental states) give rise, via the process of perceptual intuition, to perceptual belief. So, even if we accept that intuition

---

[4] Kurtz (1985) contains an extensive, critical examination of such results. Colman (1987, ch. 7) contains a brief but devastating refutation of the 'best' evidence for the existence of telepathic powers.

plays a role in sensory perception, it is puzzling how this could result in knowledge of platonic objects. Light does not bounce off such objects.

It may be argued that the perceptual process sometimes, perhaps always, gives knowledge of entities that have no causal role in the process. In the example of the tree, the causal interaction is with the front surface of a time-slice of the tree not with the tree itself, yet I come to believe and know that there is a tree in front of me, not just a slice of a tree. Intuition is posited as the process that closes the gap between the spatio-temporal slice that is caus-ally interacted with and the object which is known but not interacted with. But the known object is an object with causal powers. It has causal powers in virtue of the fact that its spatio-temporal parts have causal powers. A platonic object lacks parts with causal powers and, therefore, could not be known by such a process. Indeed, according to the usual conception of platonic objects, they have no spatio-temporal parts, since they exist outside of space and time. Even if it is accepted that we can have perceptual knowledge of an object without causally interacting with that object, but with only part of it, we still do not have an account of how we can have knowledge of platonic objects.

A somewhat different account of the involvement of intuition in sensory perception makes a more direct claim to knowledge of abstract objects. When I see a shirt, I become acquainted not only with the shirt but also with many of its properties. Suppose it is a red shirt. Then I am acquainted with its redness, and, perhaps, not only with its redness, but also with the universal redness. Now universals are abstract entities that lack causal powers, so I do not see the redness of the shirt. Rather, the story goes, I see the red shirt but intuit its redness. Mathematical structuralists, who believe that mathematics is about structures or patterns rather than individual objects, often employ such an account. (For example, Parsons (1979-80) dis-cussed later in this section.) They claim that when we see some dots on a page, we may intuit a pattern that, unlike the dots, is an acausal entity existing outside space and time.

This is just mystery-mongering. There are three possibilities. Either we (unconsciously) infer from what we perceive that platonic entities exist, in which case we need an account of how such inferences are possible. I have already discussed and dismissed the possibility of such inferential intuition. Or we directly apprehend platonic entities. Such a process is either the ESP intuition discussed above (Section 9.3) or the direct platonic intuition to be discussed below (Section 9.6). Or else some other process is being invoked, about which we are told nothing. Unless we are told more, the question of how perceptual intuition yields knowledge of platonic objects remains unaddressed. Besides, we may doubt the existence of an epistemic gap between perceiving an object and acquiring knowledge of its properties, a gap that the mysterious intuitive process supposedly closes. Is my acquaintance with the redness of the shirt any more than my coming to know that the shirt is red? And can I not come to know that the shirt is red as the result of my causal interaction with the red shirt? On the other hand, if entities like properties and patterns must be invoked, why not suppose that they have causal powers? If redness is to be part of an account of my perception of a red shirt, then it is plausible to say that it is the redness, shape, and other properties of the shirt that give rise to my perception of the shirt.

An intuitionist account of mathematical knowledge, more detailed than most, is that of Charles Parsons (1979-80 & 1983). The basis of his account is the claim that we can intuit a type when we perceive (or imagine) an object as a token of that type (1979-80, pp. 153-55). In other words, it is a variation on perceptual intuition. As such, it depends either on a causal link between the type and the perceiver, or on a mysterious *ad hoc* process to bridge the epistemic gap. Parsons seeks to close the gap by labelling the abstract types as 'quasi-concrete' objects (1983, p.26 & 1990, p. 304). This is reminiscent of, and no more convincing than, Descartes' story of the pineal gland that can be moved both by the immaterial soul and by the subtle animal spirits of the material body (1649/1969, pp. 362-63). Either these quasi-concrete objects have causal powers and there is an epistemic gap between them and what Parsons calls 'pure' abstract objects, or they have no causal powers and, therefore, cannot be given in ordinary sense perception.

9.5. COGNITIVE INTUITION

The fourth notion of intuition to be examined is what I have called cognitive intuition. The idea is that we can intuit the truth of certain propositions when we contemplate them. I do not wish to suggest that active or conscious contemplation is a necessary condition for acquiring knowledge in this way, although it does play an important role in some accounts. What is important is that it is only certain kinds of truths that can be intuitively known in this way. This distinguishes cognitive intuition from intuition-as-hunch, where there is no such restriction.

The idea that we can know that certain propositions are true simply by contemplating them is a familiar one. It usually applies to the so-called analytic and/or logical truths, such as 'All bachelors are male' and 'If p implies q, and p, then q'. Explanations as to why this is possible vary, but appeal is usually made to the fact that the propositions are true in virtue of the meanings of the terms used, or in virtue of the concepts employed, or by convention. These examples of analytic truths do not have existential import and I know of no serious claim that we can have cognitive intuitive knowledge of the existence of non-platonic objects. If it is claimed that we can have cognitive intuitive knowledge of the existence of platonic objects, then we need an explanation as to why it is not possible in the case of non-platonic objects. Either the platonic knowledge must be explained as some sort of conceptual knowledge or a special kind of cognitive intuition must be invoked which applies especially (and only) to platonic knowledge. The claim that platonic knowledge is conceptual knowledge is an alternative to the claim that platonic knowledge is intuitive knowledge and is discussed in Chapter 10. An account of platonic knowledge as conceptual knowledge might include some process of intuition, but so long as that intuition was no different than that involved in acquiring non-platonic conceptual knowledge, then it is conceptual knowledge which is doing the important work in such an account. The notion of a special platonic cognitive intuition is simply *ad hoc* and explains nothing. It won't do to claim that we have a special faculty for contemplating platonic objects, as a result of which we can cognitively intuit certain truths about them. That is just a version of direct platonic intuition (discussed in the next section), a version that incorporates perceptual intuition.

9.6. DIRECT PLATONIC INTUITION

The final notion of intuition invoked to account for platonic knowledge is the direct apprehension of platonic entities. This is usually described in terms of a faculty analogous to sense perception. Kurt Gödel (in a frequently quoted passage) expresses this notion as follows:

> But, despite their remoteness from sense experience, we do have something like a perception also of the objects of set theory, as is seen from the fact that the axioms force themselves upon us as being true. I don't see any reason why we should have less confidence in this kind of perception, i.e., in mathematical intuition, than in sense perception. (1947/1983, pp. 483-84)

It is clear that the claim that intuition is a faculty analogous to sense perception is quite distinct from the claim that intuition is part of the process of sense perception. Unfortunately, because the same term is used for both notions there is a danger of conflation. If your account of sense perception includes perceptual intuition and you also claim that we have direct platonic intuition that is a faculty analogous to sense perception, then it may be that your direct platonic intuition will include a process analogous to perceptual intuition. In fact, you may wish to claim that it is the very same process that is involved in both faculties. However, any evidence or argument for the existence of perceptual intuition will be irrelevant to the claim that we have direct platonic intuition.

Before going further, a distinction already alluded to should be noted. In the case of perception, we distinguish perception *of* objects from perception *that* something is the case. I may see the tree (perception-of) and I may see that the tree is swaying (perception-that). I may hear the bells and I may hear that the bells are ringing. The relation between these two kinds of perceiving is controversial. The usual view is that perception-of gives rise to perception-that, and that it is possible to have the former without the latter. I can see the trees without seeing that the trees are swaying or without seeing (and thus knowing) anything about the trees. However, on some accounts, perception-that is seen as more fundamental. On such accounts, only if I perceive that something is the case can I perceive objects. Whatever is the case, it is clear that we cannot have perceptual knowledge without both perception-of and perception-that.

If there really is a process of intuition analogous to sense perception, we should expect this distinction to apply to it as well. We should expect that we can have both intuitions *of* platonic objects and intuitions *that* certain mathematical claims are true. It is not clear whether Gödel is conflating or observing the distinction when he suggests that 'we do have something like a perception ... of the objects of set theory, as is seen from the fact that the axioms force themselves upon us as being true'. But exegetical concerns need not detain us. Some philosophers have claimed that we can have intuition *of* platonic objects, while others have claimed that we only have intuition-*that*.[5] Many others have not been clear on the issue. The acquisition of platonic knowledge by intuition-that alone is a variety of cognitive intuition that I discussed and dismissed in the previous section. It remains for me to argue that the knowledge of abstract objects that is required by platonism cannot be based on, or involve, a special platonic intuition-of.

There is one other possibility that should be mentioned. Some philosophers believe that propositions are intentional objects and that intuitive knowledge consists of having an intuition of such objects. For example, to know by intuition that five is a prime number is to be related in some way to

---

[5] Parsons (1979-80, p. 151) and Tieszen (1989, p. 5) argue for intuition-*of*, while Steiner (1975, p. 131) claims there is only intuition-*that*.

the proposition that five is a prime number. On such accounts, intuition-that is subsumed under intuition-of. If this is so, then my argument that platonic knowledge cannot be based on intuition-of will eliminate this possibility.

A general objection to knowledge by intuition is that intuition, whatever its exact nature, is fallible. People's intuitions concerning the same matter often differ from person to person, or vary from occasion to occasion. Intuitions are subjective and therefore cannot provide a reliable source of knowledge. Gödel responds to this criticism by saying that he does not 'see any reason why we should have less confidence...in mathematical intuition than in sense perception' (p. 484). He also suggests an analogy between the deceptions of the senses and the set-theoretical paradoxes. If the analogy between sense perception and intuition is to be taken seriously, then this is a reasonable response to the objection. If the fallibility of sense perception can be incorporated into our theory of knowledge (as it must if that theory is to be plausible), then the fallibility of intuition does not provide grounds for barring it from our theory of knowledge. Of course, if we insist on the infallibility of mathematical knowledge, then mathematical intuition would have to be in some sense an infallible process. But such a doctrine is highly implausible and finds little favour with contemporary epistemologists. Moreover, to insist on the infallibility of mathematical intuition would threaten the grip of the analogy with perception. In what way could infallible intuition be analogous with fallible sense perception?

Any subjective element of intuition should also not be a bar to incorporating it into our epistemology. Perception also has a subjective element. How we feel and what we believe and desire can affect what we perceive. If careful observers can be objective, then why not careful intuiters (such as trained mathematicians)? On the other hand, if subjectivity is an ineliminable aspect of perception, then either perceptual knowledge of an objective reality is impossible, or it is possible in spite of the subjectivity. The former is implausible and would lead to radical scepticism. To maintain the possibility of objective intuitive knowledge in the face of such scepticism would loosen the grip of the analogy with perception. If objective perceptual knowledge is possible in spite of a subjective element, then it will be because, as well as everything else that is going on, the right sort of relationship exists between the perceiver and an objective external reality. Gödel makes a similar claim for intuition:

> It by no means follows, however, that [intuitions], because they cannot be associated with actions of certain things upon our sense organs, are something purely subjective... Rather they, too, may represent an aspect of objective

reality, but, as opposed to the sensations, their presence in us may be due to another kind of relationship between ourselves and reality. (p. 484)

Unfortunately, it is the positing of this 'other kind of relationship' which threatens to undermine the case for knowledge by intuition. What is the nature of this relationship? All we have to go on is the analogy with perception. But a distinguishing feature of the relationship between the knower and reality in the case of perception is causation. Perception is a causal process. It is puzzling how a process could be analogous to perception and yet not involve causation. Furthermore, no independent evidence is offered for our having a faculty of intuition, or even for recognizing when we are exercising it. Talk of this mysterious faculty of intuition is tainted with an air of occult mysticism. The notion is *ad hoc* and the analogy 'broken-backed' (Hale 1987, p. 79) or 'flabby' (Chihara 1982, p. 217).

These criticisms may be unfair if Gödel's analogy is intended in only a restricted sense. One suggestion is that the common feature between sense perception and mathematical intuition that Gödel is drawing our attention to is that in the case of sense perception we have no choice as to what seems to be the case, and the same is true when we contemplate the axioms of set theory. As Berkeley puts it:

> When in broad daylight I open my eyes, it is not in my power to choose whether I see or no, or to determine what particular objects shall be present to my view: and so likewise as to the hearing and other senses; the ideas imprinted on them are not creatures of *my* will. (1710/1965, p. 72)[6]

Is Gödel simply pointing out that we have the same lack of choice with respect to the axioms? If so, then we are no closer to an account of *how* platonic knowledge is acquired by intuition. Or perhaps his comments should be seen as offering evidence for the existence of another process for acquiring knowledge of objective, external reality. But if it is evidence for anything, it is evidence for a *causal* process, especially in the light of his metaphor of the axioms 'forcing' themselves upon us as being true.

Besides, it is not at all clear that the cases are similar. Perhaps there are axioms that we feel bound to accept. One such may be the axiom of extensionality which states that two sets are equal if and only if every element of one is an element of the other, and vice versa. But other axioms of set theory such as the axioms of choice, infinity, and reducibility have been hotly debated. Also, recall Cantorian set theory with the Continuum Hypothesis (CH) as an axiom and non-Cantorian set theory with its negation

---

[6] Of course, Berkeley concludes that the source of the ideas is the will of God, rather than material objects, but the relationship is with an objective reality for all that.

(~CH) as an axiom (Section 5.5). Neither CH nor ~CH 'forces' itself on us as being true. Gödel concedes as much when he talks of verifying an axiom by other means, such as by examining what can be proved with its help and what cannot (1947/1983, pp. 477 & 485). Finally, and crucially, axioms with existential import do not have this feature, especially when they are given a platonistic construal. I, for one, do not feel bound to accept that the empty set exists independently as an acausal object. This very book bears witness to that.

James Brown argues that the mysteriousness of the purported process of intuition should not be a bar to our accepting Gödel's explanation. He suggests that sense perception is equally mysterious:

> In the case of ordinary visual perception of, say, a teacup, we believe that photons come from the physical teacup in front of us, enter our eye, interact with the retinal receptors and a chain of neural connections through the visual pathway to the visual cortex. After that we know virtually nothing about how beliefs are formed. The connection between mind and brain is the great problem of the philosophy of mind. Of course, there are some sketchy conjectures, but it would be completely misleading to suggest that this is in anyway 'understood'. Part of the process of cognition is well understood; but there remain elements which are just as mysterious as anything the platonist has to offer. (1990, p. 108)

But the objection to intuition is that it is *wholly* mysterious. Brown himself supplies sufficient detail about perception to show that the process is not wholly mysterious. There may be gaps in our understanding, but there are parts of the process that we do understand. Continuing scientific investigation is gradually filling in those gaps, giving us no reason to suppose that our overall picture is radically mistaken. The mystery of direct platonic intuition goes deeper than this. When investigating an unknown causal process, scientists have a fairly clear idea of what sort of hypotheses may appropriately be tested. Whatever the outcome of their tests, further hypotheses will be suggested. But in the case of investigating a non-causal process, it is not even clear what, if any, hypotheses are appropriate. An analogy between the mysteriousness of perception and the mysteriousness of intuition cannot be sustained.

Brown tries another line of attack. He suggests that the correctness of our theory of perception is irrelevant to whether or not we can have perceptual knowledge. He points out that our ancestors' beliefs about perception were quite wrong and that it is possible that our beliefs about it are equally wrong.

> Nevertheless, whether we are right or wrong in our account of perception, we are (and our ancestors were) rightly convinced that there are material objects such as trees and tables which exist independently of us and that they

somehow or other are responsible for our knowledge of them. Whether we
have the right theory, the wrong theory or no theory at all, it is reasonable to
reject Berkeley['s idealism]. (p. 109)

Brown's argument seems to be that the perception/intuition analogy
amounts to no more than the claim that *if* we have faculties of perception and
intuition, then we can use those faculties to obtain knowledge of the external
world, irrespective of what theories we have about those faculties. I shall not
dispute this conditional claim (see Section 8.2). The same claim is true of
faculties of precognition, clairvoyance, and religious mysticism, *if* we have
such faculties. But the claim is worthless as a response to the request that
platonists explain how we acquire knowledge of the existence of platonic
objects.

Before leaving direct platonic intuition, I turn to two bold, but
contrasting, responses to the objection that the analogy between perception
and intuition is quite unhelpful because one is a causal process while the
other is not. Penelope Maddy (1980 & 1990) offers the bold conjecture that
the analogy holds good because both are causal processes. She claims that
sets of physical objects are located in space and time, and that we can
perceive them just as we can perceive physical objects, for example, by
looking at them and seeing them. Maddy is concerned only with sets because
she believes that all of mathematics can be reduced to set theory. So,
according to Maddy, sets are entities with causal powers. There are a number
of problems with Maddy's position but I shall not discuss them here. Her
brand of 'physicalistic platonism' does not assert that we can have knowledge of the existence of acausal objects, so it is outside the scope of this
book.

The mirror image of Maddy's position is Richard Tieszen's claim (1989)
that Gödel's analogy is appropriate because neither mathematical intuition
nor sense perception is a causal process. His claim is based on Husserl's
phenomenological accounts of the two processes. Husserl's account of sense
perception is mostly concerned with an analysis of what he calls perceptual
intuition. He gives a similar, but less developed, analysis of mathematical
perception. Tieszen first develops an account of mathematical intuition along
phenomenological lines and then argues for the analogy between the two
kinds of intuition. According to the phenomenological view, perceptual intuition is thought of:

> in terms of sequences of indexical partial perceptions of objects and not in
> terms of some sort of special relation, causal or otherwise, and independent of
> this process, which is supposed to put us in touch with real objects' (pp. 63-
> 64).

The emphasis is on intentional cognitive acts and the contents of those acts. The objects towards which those acts are directed are 'bracketed' (as the phenomenological jargon has it), and play no part in the analysis. The reason for this seems to be that the object may or may not exist and '[w]hether the object exists or not depends on whether we have evidence for its existence, and such evidence would be given in further acts carried out through time' (p. 23). That this view of perception entails idealism seems inescapable. There is no clear distinction between what is perceived and the perception of it, or between actual existence and evidence for existence. If the existence of an object *depends* on our having certain evidence, and that evidence, in turn, depends on our performing certain acts of perception, then it is difficult to see in what sense such an object could be mind-independent.

Tieszen provides further evidence for the idealism implicit in this view when he states that:

> The point of phenomenological reduction is just that we are to investigate the structure of acts and sequences of acts, and to do so does not require considerations about causality. [W]e are not committed to providing an analysis of the condition that the state of affairs referred to by S *causes* M to believe that S... On our view there is an analogy between mathematical and perceptual intuition and an analysis of any condition about causes is not involved in either case. It should be noted that we are not saying that ordinary perceptual objects are not given as objects to which we are causally related, or as objects that may be causally related to one another. On the contrary, they would certainly be recognized as so given, unlike mathematical objects. (pp. 179-180)

No doubt, the pink rats that the inebriate hallucinates are *given* as objects to which he and other objects are causally related. But they are not actually causally related because they do not actually exist. This is one reason why the inebriate does not *know* that there are pink rats in front of him. It may be that the Husserl/Tieszen analysis provides a splendid account of what goes on when someone has a perceptual or mathematical 'intuition', and it may be that those two processes are analogous. But the analysis does not provide an account of processes by which we acquire knowledge of external reality. The analysis of perceptual intuition is insufficient as an account of how we acquire perceptual knowledge since it lacks the crucial link of causation, and this insufficiency makes it useless as an analogy for a process by which we acquire knowledge of platonic objects.

I conclude that, although there may be actual faculties or processes to which the term 'intuition' may be appropriately applied, none of them can plausibly account for platonic knowledge. To posit another sort of intuition especially for platonic knowledge would be *ad hoc* and, therefore, cannot help to explain how it is that we can have knowledge of the existence and nature of platonic objects. More than a century later, Frege's warning should still be heeded. 'We are all too ready to invoke inner intuition, whenever we cannot produce any other ground of knowledge.' (1968, p. 19)

CHAPTER 10

APRIORISM

10.1. *A PRIORI* PLATONIC KNOWLEDGE

There is a long tradition of regarding mathematical knowledge as *a priori* knowledge. But most detailed accounts in this tradition are not overtly platonistic and many are clearly not. In this chapter I examine three recent accounts that explicitly combine the claims that mathematical objects are platonic and that we can know *a priori* that they exist.

10.2. PLATONIC KNOWLEDGE AS CONCEPTUAL KNOWLEDGE

The first account is the claim is that platonic knowledge is conceptual knowledge. Conceptually knowing that *p* is usually understood as knowing that *p* simply as a result of understanding that *p*. It would be surprising if knowledge of the existence of platonic entities (or any entities, for that matter) could be obtained in this way. There have been no successful ontological arguments for the existence of God, and the reasons for this are well understood. Conceptual knowledge about God is knowledge of the properties that God must have, *if God exists*. No matter how clear our understanding of what it means for God to be omnipotent, omniscient, omnibenevolent, or maximally great in every possible world, this understanding cannot tell us whether or not an entity exists with those properties.

Suppose *q* is a statement that asserts or entails the existence of some platonic object **b**. Now to understand *q*, one either already needs to know that **b** exists or one does not already need to know. If one already needs to know that **b** exists, then one would already have to know of **b**'s existence by a non-conceptual process or by a conceptual process. Prior knowledge of **b**'s existence by a non-conceptual process would mean that q is not known by a purely conceptual process. On the other hand, prior knowledge of **b**'s existence by a conceptual process results in a vicious circle, since to know that **b** exists by a conceptual process would require understanding '**b** exists' but, *ex hypothesi*, this would require one's already knowing that **b** exists.

If one does not already need to know that **b** exists, then we are left wondering how understanding alone could yield existential knowledge. The usual examples of facts that can be known by such a process do not entail any existence claims. For example, 'all bachelors are male' and 'all groups contain a unit element'. Apprehending the full import of some concept is one thing, knowing that the concept has any instances is another. There are good arguments to the effect that the former can be known conceptually. We lack an argument to the effect that the latter can also be so known.

Crispin Wright (1983) offers an argument based on Frege's argument for platonism. The logicist programme that was inspired by Frege's work is widely acknowledged to have failed, so it will be sufficient for my purposes to concentrate on Wright's argument.[1]

Wright's argument may be summarised as follows:

(W1)   If a range of expressions function as singular terms in true statements, then there are objects denoted by those expressions.

(W2)   Numerical expressions do function as singular terms in many true statements (of pure and applied mathematics).

∴ (W3)   There exist objects denoted by those numerical expressions (i.e. there are numbers).[2]

Premise (W2) may be considered as two separate claims:

(W2a)   Numerical expressions function as singular terms in many statements (of pure and applied mathematics).

(W2b)   Many of those statements are true.

---

[1] My discussion draws on the criticism of Wright's argument by Field (1989, ch. 5) and by Musgrave (1986).

[2] Based on Hale's version of the argument (1987, p. 11). It omits details concerning syntactic and semantic functioning, and Frege's context principle, which I do not wish to dispute.

Wright offers a range of argument and evidence in support of (W1) and (W2a). Suppose we provisionally accept (W1) and (W2a). The crucial epistemological question is now 'How do we know that some statements (of pure and applied mathematics) are true?' After all, they refer to objects from which we are causally isolated.

Wright's answer is that they are equivalent to other statements whose truth we can verify 'by ordinary criteria' (1983, p. 14). For example, consider the following pair of statements:

(a)   There are exactly three apples in the house.

(b)   The number of apples in the house is three.

We can know the truth of (a) by counting apples, a process that does not involve us in causal contact with numbers. If we know that (a) and (b) are equivalent, then we can know that (b) is true. From (b), we can infer that at least one number exists.

But how do we know that (a) and (b) are equivalent? That depends on what sort of equivalence they have. Are they logically equivalent? Two statements are logically equivalent if, when translated into the language of an appropriate formal system of logic, they are inter-derivable. This does not appear to be the case and Wright agrees.

Wright believes that they are conceptually equivalent (1983, pp. 151-52). Statements are conceptually equivalent if the identity of their truth-values follows logically from a conceptual truth. For example, 'John is a bachelor' is conceptually equivalent to 'John is an unmarried man' because they are mutually deducible via the conceptual truth 'Bachelors are unmarried men'.

What conceptual truth, according to Wright, guarantees the equivalence of (a) and (b)?

(N=)   The number of things that are A is the same as the number of things that are B if and only if there are exactly as many As as Bs. (p. 104)

The claim is that only by understanding and accepting something like (N=), can we acquire the concept of number. But if we accept (N=), then whenever we acquire knowledge of statements like (a) we cannot rationally avoid coming to know that numbers exist.

The problem with (N=) is that it has existential import. It is trivially true that there are exactly as many Fs as there are Fs. Hence, by (N=), the number of things that are F is the same as the number of things that are F. From which we can deduce that something exists which is the same as the number of things that are F.

It is a conceptual truth that conceptual truths do not have existential import. So (N=) is not a conceptual truth. On the other hand:

(N=*)   If numbers exist, then (N=)

is a conceptual truth. Unfortunately, (b) is not deducible from (a) via (N=*). Only by assuming that we already know that numbers exist can Wright's argument show how we can know that numbers exist. Not surprisingly, Wright's ontological argument for the existence of platonic objects is no more convincing than are any of the ontological arguments for the existence of God.

(a) and (b) are neither logically nor conceptually equivalent, but there is the nagging feeling that it sounds absurd to deny that they have the same truth value. So what sort of equivalence do they have? Perhaps their equivalence lies at a deeper level than of logic or concepts. Could reality be such that at is simply impossible that (a) and (b) could have different truth-values, and that we recognise this even though it cannot be captured by our talk about logic or concepts? This assertion of metaphysical equivalence is a strong claim and requires a strong argument. It may be that Hale (1994) has something like this in mind. I discuss this in the Section 10.4.

Hartry Field bites the bullet and declares that they are not equivalent because (a) is true but (b) is false, the latter being false because there are no numbers, although he agrees that if numbers did exist, such statements would be equivalent (1989, pp. 159-60). A less dogmatic position could be that the truth of (b) is unknown (and perhaps unknowable), because we do not know (and perhaps cannot know) whether numbers exist.

On the other hand, Alan Musgrave allows that both (a) and (b) are true, but only if they are seen as idiomatically equivalent, not conceptually equivalent (1986, p. 107). Here (b) is idiomatic (or a *façon de parler*) for (a), just as:

(i)   Jenny Shipley gets my goat

is idiomatic for:

(ii)   Jenny Shipley upsets me.

Although both (i) and (ii) are true, it does not follow that something exists which is my goat. Just as 'my goat' does not function semantically as a singular term in (i), 'three' does not function semantically as a singular term in (b).

It is not necessary to adjudicate between Field and Musgrave. It is quite natural to say of (i) that it is not literally true, but that it is true in a manner of speaking, and we can say the same of (b), at least until such time as we can satisfy ourselves that numbers exist. Putting it another way, (N=) is useful, not as a means of gaining access to platonic knowledge, but as a means of paraphrasing away existential number-talk.

10.3. REALISTIC RATIONALISM

Jerrold Katz (1998) offers his Realistic Rationalism as a philosophical position that integrates the claim that mathematical objects are abstract with the claim that we have non-empirical knowledge of those objects. To meet the epistemic challenge to platonism he offers a rationalist epistemology. In offering his epistemology, it is clear that Katz takes the epistemic challenge seriously. He accepts that it is the task of the platonist

> to provide an account of [the] other kind of relationship that explains how we come to stand in that relationship to the realm of abstract objects and, with no window on that realm, come to know what things are like there. (p. 34)

In return, Katz demands that Benacerraf's semantic challenge for anti-platonism be taken seriously. On the face of it, the sentences:

(1)   Seventeen is a prime number

(2)   Al Gore is a politician

are grammatically on a par and should be treated as such. Katz argues that in cases such as:

(3)   John is easy to please

(4)   John is eager to please

it is appropriate to treat them differently because their 'surface similarity masks deep grammatical difference' (p. 30). 'John' is the object of the infinitive in (3) but the subject in (4). The different grammatical treatment is legitimate because it is grammatically driven. On the other hand, Katz suggests that to treat (1) and (2) differently in order to avoid a commitment to platonism is to be driven by a 'linguistically irrelevant' aim.

This argument is too strong. There are non-grammatical reasons for treating (2) differently from the structurally similar:

(5)   James Bond is a spy.

Although those reasons may not be ultimately convincing, they cannot be dismissed out of hand as linguistically irrelevant. Besides, not all anti-platonists would wish to treat (1), (2) and (5) differently. A Fieldian approach would have it that they all have the same deep grammatical structure, but that (1) and (5) are simply false, albeit that (1) and (5) have closely related sentences that are true:

(1′)   According to standard arithmetic, seventeen is a prime number.

(5′)   According to Fleming's fiction, James Bond is a spy.

The above grammatical considerations lead Katz to the conclusion that mathematics is about abstract objects. He also accepts that mathematical truths must be necessary truths, apparently on the grounds that Quine's arguments that there are no necessary truths fail. This leads to the conclusion that 'we cannot know truths about mathematical objects in the same way that we know truths about natural objects' and that 'the existence of mathematical knowledge shows that there must be a different way of knowing mathematical truths' (p. 32). It is with these conclusions firmly in place that he sets about his task of providing a rationalist epistemology or, more precisely, an account of our *a priori* knowledge of abstract objects.

Katz summarises his account as follows:

> [I]nvestigation in the natural sciences seeks to *prune down the possible to the actual*, while investigation in the formal sciences seeks *to prune down the supposable to the necessary*... Since pruning down the supposable to the necessary requires only reason, formal knowledge is *a priori* knowledge. Since pruning down the possible to the actual requires interaction with natural objects as well as reason, natural knowledge is *a posteriori*. (p. 59, his italics)

He starts by explaining why he believes that 'the entire idea that our knowledge of abstract objects might be based on perceptual contact is misguided' (p. 36). Since natural objects can be otherwise than they actually are, contact with them enables us to discover how they actually are. But abstract objects cannot be other than they are. They have all their intrinsic properties necessarily. Perceptual contact can provide us with information as to how things are (as opposed to how they might have been), but it cannot provide us with information as to how things must be. So, he concludes, perceptual contact cannot ground beliefs about abstract objects.

This is puzzling. I have a pet called 'Bijou'. You may wonder whether Bijou is a dog or a cat or a goldfish or what. Perceptual contact will establish that she is a cat. Suppose being a cat is an essential property of Bijou. Katz apparently accepts that natural objects may have essential properties (p. 36). Bijou could not be other than a cat. But this does not prevent you from coming to know that she is a cat by perceptual means. So why shouldn't one be able to discover that, for example, 17 is a prime number by perceptual contact? It is not the fact that 17 is necessarily prime that prevents this, but its acausality. You may object that you cannot know via perception that Bijou *must* be a cat. Perhaps so. But note that Katz already assumes that abstract objects have their properties necessarily before advancing his argument. Suppose I already know that 17 has its properties necessarily. If I could then observe that 17 is prime, I could conclude that 17 must be prime. Furthermore, even if we grant that mathematical truths are necessary truths, it does not follow that all mathematical knowledge is necessary knowledge (in the sense that one knows that certain truths are necessarily true). Why couldn't a mathematical novice learn that 17 is prime without learning that 17 is necessarily prime? Katz's argument does not establish that perceptual contact (supposing it to be possible) could not play a role in the acquisition of knowledge about abstract objects.

Never mind. Katz does not need this strong claim as part of an account of how we obtain *a priori* knowledge of abstract objects. He continues his account by offering *reductio* proof as a paradigm method for acquiring mathematical knowledge by reason alone. As an example he presents a proof that two is the only even prime. The supposition that there is another even prime gives rise to a contradiction. Such a proof, he claims, 'provides us with adequate grounds for knowledge of a proposition about abstract objects by showing that it is impossible for the objects to be other than as the proposition says they are' (p. 40).

Objections crowd in. There is nothing special about the content of *reductio* proofs. *Reductio* proofs are instances of a particular form of argument and, as such, may be arguments about anything. If *reductio* proofs can provide us with knowledge of their conclusions, then they can provide us with knowledge of contingent facts about concrete objects, since the conclusion of a *reductio* proof may be any proposition. However, we do sometimes speak as though such conclusions tell us how things must be with the objects that they are about. Consider those brain-teasers in which we are told various things about certain persons, their professions and their neighbours, and invited to work out which of them is the Baker. After a careful deduction (perhaps employing a *reductio* proof) we may announce that Mr Smith *must be* the Baker, but we would not thereby be claiming that no matter what the world were like it would be inhabited by Mr Smith the Baker. Even if the premises of the proof are true, it is not a necessary truth that Mr Smith is the Baker.

Indeed, we can construct a proof concerning contingent, concrete objects that closely parallels Katz's proof that two is the only even prime. Consider a line of soldiers in which every second soldier from one end is wearing a red coat and every other soldier is wearing a blue coat. Also, every second soldier, apart from the second soldier in the line, is wearing a hat, as is every third soldier apart from the third soldier in line, every fourth soldier apart from the fourth, and so on. All other soldiers are hatless. Call the second soldier in line 'Jerry'. The following argument proves that, no matter how long the line, Jerry is the only hatless soldier wearing a red coat.

We see that Jerry is wearing a red coat and no hat. Supposing that another soldier is wearing a red coat but no hat, that soldier must come either before or after Jerry. If the soldier is before Jerry, it has to be the first soldier, but then he is wearing a blue coat, not a red one. But if the soldier comes after Jerry and is wearing a red coat, then he is one of every second soldier and hence is wearing a hat. Since the law of trichotomy cannot fail here, there is no soldier wearing a red coat and no hat other than Jerry.

Replace the soldiers with numbers, Jerry with the number two, red coats with evenness, blue coats with oddness, and hatlessness with primeness and we have Katz's proof that the number two is the only even prime. But this gives us no more reason to believe that numbers exist than that these soldiers exist.

Katz claims that 'a proof of a proposition P that is not itself a modal statement establishes the modal statement "Necessarily, P" in virtue of the fact that it is a proof of P' (p. 40). This claim is simply false. It commits the fallacy of misconditionalisation. The proof only establishes, "Necessarily, P follows from the premises." (See Stove 1972.) Katz seems to be aware of the problem. He attempts to retrieve the situation by claiming that for his example (the proof that two is the only even prime) the premises 'cannot be faulted'. In other words, he claims that the premises are necessarily true. I do not dispute (nor should anyone else) that any valid argument with necessarily true premises has a necessarily true conclusion. What is at issue is how we *know* that those premises are true!

The premises of Katz's proof are presumably the Peano axioms, or something akin to the Peano axioms. The Peano axioms assert that the natural numbers exist and are related to each other in certain ways. Platonists agree that these numbers do exist and that they are acausal objects. A platonist epistemology must tell us how we can come to know that they exist. As a bonus it might tell us how we can come to know that they exist necessarily. An *apriorist* platonist epistemology must tell us how we can come to know these things *a priori*. Katz's 'Outline of a Rationalist Epistemology' starts on p. 34 of his (1998) and is complete by p. 60. Nowhere on those pages does he address these crucial issues. His epistemology is not a platonist epistemology. It is a rationalist epistemology of sorts. It tells us how we can have *a priori* knowledge of certain necessary truths. Essentially, it tells us how we can know *a priori* that an argument is valid. In other words, it tells us how we can know *a priori* that a conclusion necessarily follows from certain premises. I agree that we can know such things *a priori*. Only a dyed-in-the-wool extreme empiricist would deny it.

So Katz's platonist epistemology is all a bit of a muddle. It promises much but fails to deliver. He starts by promising to tell us how we can have *a priori* knowledge of a platonic realm. This slides into an account of how we can have *a priori* knowledge that a mathematical truth is a necessary truth. But what we finish up with is no more than an account of how we can have *a priori* knowledge that an argument is valid. Katz defends this account against a number of objections, but since they are only objections to his final account, which I do not need or wish to dispute, there is no need to discuss them here.

10.4. APRIORITY AND NECESSITY

Bob Hale (1994) argues that platonism is not epistemologically bankrupt. Interestingly, he starts more or less where Katz leaves off. Unfortunately, he then comes full circle and makes more or less the same mistake that Katz does. Hale believes that a platonist epistemology must provide an account of how we acquire knowledge of necessary truths, since, on his version of platonism, mathematical truths are necessary truths. He also acknowledges that some of the claims of mathematical platonism are existential claims. Hale's strategy is an attempt to turn the tables on his anti-platonist opponents. He notes that most anti-platonists (Hartry Field is his particular target) agree that we can have knowledge of necessary truths. Typically, anti-platonist theories of mathematics assert that accepted mathematical statements are logical consequences of certain axioms, or that certain mathematical theories are consistent. To know that such assertions are true is to have knowledge of necessary truths. (Hale accepts that his argument has little purchase against anti-platonists who are sceptical of necessary truth.)

Hale's thesis is that once we allow that knowledge of necessary truths is possible, then there is no *special* epistemological problem with mathematical statements construed platonistically (1994, p. 300). The idea is that whatever account is to be given of our knowledge of necessary truth it will not involve a causal connection between the belief that p and the fact that p. This is because such a connection appears neither possible nor necessary for knowledge of logical consequence. Hale's point appears to be a reasonable one. Note that the requirement of k-causal connection is so weak that it is met in the case of a belief in a necessary truth no matter what the cause of the belief.

The difficulty for Hale is that his mathematical platonism requires that mathematical existence claims be both literally and necessarily true. I have already rejected the notion of necessary existence, as have many others. Hale considers such a rejection to be too hasty. He maintains that the failure of ontological arguments for the necessary existence of God is irrelevant, since mathematical necessity and arguments for that necessity may be of a quite different variety. He argues that the ontological arguments 'can, and should, be faulted on grounds which do nothing to call into question the general idea that there may be necessary existential truths' (p. 318).

The onus is now on Hale to provide an argument for the necessary existence of mathematical entities. His argument, like Crispin Wright's, draws on Frege's logicist argument for mathematical platonism. At the nub of his argument is a derivation that demonstrates that the necessary existence of the number zero (and, hence, the necessary existence of all the cardinal numbers)

> straightforwardly follows from Frege's criterion of identity for cardinal numbers—that is, that for any F and G, the number of Fs is identical with the number of Gs iff the Fs and Gs are in one-one correspondence—together with the supplementary premise that no object is self-distinct. Since both premises are necessarily true, so is the conclusion. (p. 324)

Hale offers no argument for the necessary truth of Frege's criterion, which is essentially Wright's (N=). Since (N=) entails the existence of numbers, he effectively assumes what he purports to prove.[3]

I conclude that none of the above provides a convincing account of how we can know *a priori* that platonic objects exist.

---

[3] Janet Folina (2000, p. 327) makes a similar point.

CHAPTER 11

INDISPENSABILITY AND PLATONIC KNOWLEDGE

11.1. POSTULATION AND INDISPENSABILITY

Willard Quine and Hilary Putnam argue that the methods by which we confirm scientific theories are the means by which we acquire knowledge of platonic objects. In outline, they argue as follows.[1] Our best theories about the world postulate entities that we cannot observe (for example, electrons) in order to make sense of our experiences. But those same theories postulate platonic objects (for example, numbers and sets). Mathematical objects are just as indispensable to science as theoretical entities like electrons. Electron theory quantifies over numbers and other platonic entities, just as it quantifies over electrons. So we have the same reason for thinking that numbers exist as we do electrons. Our knowledge of platonic objects is obtained in the same way as our knowledge of physical objects, from sense experience. The process by which we obtain platonic knowledge is the same as the process by which we obtain non-platonic scientific knowledge. Indeed, platonic knowledge is an indispensable part of scientific knowledge.

Alan Musgrave and Elliot Sober have challenged the claim that platonic objects are in the 'same epistemological boat' as physical objects. Musgrave (1986, pp. 90-91) points out that we can imagine obtaining evidence that makes it doubtful that electrons exist, but we cannot imagine what sort of evidence would count against the existence of numbers. If numbers exist, then they exist necessarily. Unlike electrons, there is no empirical evidence that can count for or against their existence. Sober (1993) points out that

---

[1] Quine (1961, ch. 1 & 1966, ch. 20) and elsewhere; Putnam (1979).

156

Musgrave's argument appears to depend on what we can and cannot imagine, which is somewhat indeterminable and variable. He also disagrees with the suggestion that we cannot have empirical evidence for entities that exist necessarily. (I suggest that there is also a problem with the supposition that if numbers exist, then they exist necessarily.) Sober advances a variation on Musgrave's argument, seeking to avoid Musgrave's problematic suppositions. But he argues from the perspective of his 'contrastive-empiricist' theory of scientific confirmation, in opposition to Musgrave's Popperian perspective.

Drawing on Musgrave and Sober, I propose the following argument that is neutral with respect to any of the standard theories of scientific confirmation. Suppose we have a number of competing theories $T_1, T_2, ..., T_n$, and, after appropriate observations and experiments, theory T_k is confirmed as the best theory, the theory that should be believed. If T_k entails M, then we should also believe M. It may seem that M has now also been empirically confirmed. But suppose (as will be the case if M is part of mathematics) that for each T_i, T_i entails M.[2] Then M has not been confirmed by the observations at all, since it would have survived as part of the best theory, whatever those observations had turned out to be. To test (and hence to confirm) M empirically, it must be tested against a theory that entails ~M. But what sort of observations would favour M over ~M? Even if we could devise such an experiment, the important point is that *no such experiment has been performed* and, therefore, mathematical knowledge has not been acquired by this method.

Electrons and platonic objects do appear to be in different epistemological boats. If we have platonic knowledge, it is not acquired in the same way as scientific knowledge of physical posits. So the question as to what process yields platonic knowledge has not been answered. But the Quine/Putnam argument suggests a further problem for the anti-platonist. What is the status of our best theory T_k? If M is embedded in T_k, in such a way that we cannot express T_k without its entailing M, then it seems that we cannot believe T_k without believing M. It is as though T_k is the ontological cargo we want on board, but it cannot be loaded without also allowing on board the platonic rats that infest it. And we cannot jettison the vermin without losing the cargo as well. Perhaps electrons and platonic objects *are*

---

[2] Sometimes, in physics, there have been competing theories that differ in their mathematical commitments. But the mathematics to which they are respectively committed can be reduced to set theories, which have in common a hierarchy of sets at least up to cardinality $\aleph_0$.

in the same epistemological boat, but came aboard by different means, or for different reasons.

This suggests an alternative line of reasoning for platonists. The Quine/Putnam argument can be put independently of assumptions about empirical confirmation. It does not need the claim that platonic and non-platonic objects are known in the same way. Indeed, the version of the indispensability argument that Hartry Field regards as the best version avoids just those things.

> [T]he theories that we use in explaining various facts about the physical world not only involve a commitment to electrons and neutrinos, they involve a commitment to numbers and functions and the like... [T]he very same explanations in which the postulation of unobservables is essential are explanations in which the postulation of mathematical entities is essential: mathematics enters essentially into our theory of electrons... There seems to be no possibility of accepting electrons on the basis of inference to the best explanation, but not accepting mathematical entities on that basis...for the very same explanations are involved in both cases. (Field (1989, pp. 16-17)

The main assumption underlying this argument is that whatever scientists accept as the best theory is what we are most justified in believing, and surely this is right. On the basis of this argument, platonists can describe a process by which we acquire platonic knowledge, the 'indispensable-postulate' process. Being an indispensable part of a well-confirmed theory is sufficient for a postulation to become an item of knowledge. Some posits of the theory will have been empirically tested (otherwise the theory will not be well-confirmed), but others will not. Those other posits are known, none-theless.

Some platonists will reject this account of platonic knowledge on the grounds that being an indispensable part of a scientific theory is not sufficient for knowledge. They will argue that only if the mathematics is known prior to its inclusion in the theory, can the theory as a whole be known. If a scientific theory contains parts which have not been empirically tested or which are not known by another, prior method, then the theory as a whole cannot be known. Of course, such platonists will need to claim that there is another such method available. So this objection to 'indispensable-postulate'-ism can give no comfort to anti-platonists.

11.2. FIELD'S PROJECT

An obvious line of attack for anti-platonists is to question the indispensability of mathematics to our best, current, scientific theories. This is the approach of Hartry Field (1980 & 1989). He argues that the indispensability argument 'can be undercut if we can show that there are equally good theories and explanations that don't involve commitment to numbers and functions and the like' (1989, p. 17). That this may be feasible stems from what Field calls the 'conservativeness' of mathematics. Mathematical theories should not only be logically consistent but consistent with any (consistent) scientific theory. Platonists who believe that mathematics is necessarily true must agree that mathematics is conservative in this sense, because conservativeness is entailed by necessary truth. It follows from the conservativeness of a mathematical theory M that, for any consistent theory S, any conclusion C about the physical that which follows from S & M, also follows from S alone.

To see that this is so, suppose C follows from S & M, but not from S alone. Then S & ~C is consistent and so, by assumption, M is consistent with S & ~C. Hence, S & M is consistent with ~C, but then C could not follow from S & M, contrary to assumption.

Now if S is a non-platonistic theory (i.e. it is free of any platonistic existential import), then we can see that a conservative theory M could be very useful for drawing conclusions from S (it may make the deductions much shorter). However, we could use M in this way without its being true and, hence, we would not need to accept M. In many of its scientific applications, mathematics is a useful, but (in principle) dispensable, aid to deduction. On the other hand, if S does have platonistic existential import (in particular, if its axioms are platonistic) but can be shown to be statable in non-platonistic terms, then once again M will be useful but dispensable for deriving non-platonistic conclusions from S (Field 1980, ch. 1). Field gives details of a number of cases where mathematical applications are dispensable to science (ch. 2-8).

A platonistic theory M will be indispensable to a theory T only if it is embedded essentially in T. To say that M is embedded essentially in theory T is to say that T entails M and that T cannot be expressed as T* + M, where T* (which is non-platonistic) does not entail M. Field admits that 'we do not know in detail how to eliminate mathematical entities from every scientific explanation we accept' (1989, p. 17). Whether he has done enough to show that it can, in principle, be done is the subject of strong debate. It is argued that the cases he has dealt with do not successfully avoid a commitment to platonism and that there are cases he has not yet dealt with (for example, quantum mechanics) for which there is good reason to believe that 'nominalisation' is impossible.[3] Rather than discuss these objections, I argue that 'indispensable-postulate'-ism may be rejected for reasons that are independent of the success of Field's project.

11.3. REVISIONIST MATHEMATICS

An alternative line of attack for anti-platonists is to question the claim that our best scientific theories are committed to the existence of platonic entities. One way of doing this is to accept that mathematical claims (whether embedded in scientific theories or not) are true, but to argue that they do not commit us to a platonist ontology. There is a long tradition of attempts to provide a non-platonistic construal of mathematical statements. The tradition goes back at least to John Stuart Mill, and probably further. Some recent attempts have been advanced by Charles Chihara, Geoffrey Hellman, and Philip Kitcher. According to Chihara's modal-constructivism (1990), mathematics can be interpreted as being about the possible construction of (tokens of) open sentences. According to Hellman's modal-structuralism (1989), mathematical claims may be interpreted as asserting no more than the possibility of there being certain mathematical structures. According to Kitcher's empiricist account (1984), mathematics is about the possible operations of an ideal agent.

Such 'revisionist' theories have been subjected to a number of criticisms. One criticism is that it is implausible to suppose that in order to reveal what mathematical knowledge is about, we must 're-express that knowledge in forms of language never employed by its possessors' (McFetridge 1985, p. 323).[4] Could it be that for centuries mathematical claims have been made by people who had little or no idea as to what they were claiming? This criticism seems to be based on the assumption that people, mathematicians in particular, have always been clear about the ontology of mathematics. Isn't it impertinent of latter-day philosophers of mathematics to suggest that mathematicians have always meant something other that what they took themselves to mean? But there is plenty of evidence that this is indeed the case. Ask a mathematician if there is a prime number greater than 365 and you will get an affirmative answer. Enquire as to what sort of an object it might be and you are likely to get confusion. Even those who are not confused will

---

[3] For example, Shapiro (1983 & 1993, pp. 458-65), Liston (1993), Chihara (1990, ch. 8), and many others. But see also Mundy (1992) and Balaguer (1998b, ch. 6) for some recent attempts to nominalise quantum mechanics.

[4] See also Burgess (1990) p. 7.

differ among themselves. It has been suggested that the mathematicians of the world are 65% platonists, 30% formalists, and 5% constructivists. On the other hand, there is the suggestion that mathematicians are more often than not divided within themselves; that they are platonists on weekdays and formalists on Sundays (Davis & Hersh 1981, pp. 321-22).[5] This being the case, it is not unreasonable for philosophers of mathematics to suggest a revised account of mathematical knowledge that is, by their lights, ontologically and epistemologically 'squeaky-clean'.

More telling criticisms are based on the claim that the revisionist accounts are not as ontologically or epistemologically pure as their proponents suppose, or that they do not capture everything that is of mathematical value. But my rebuttal of postulationism will not depend on the success of any particular revisionist programme.

11.4. VARIABLE ONTOLOGICAL COMMITMENT

Another way of questioning the claim that our best scientific theories are committed to the existence of platonic entities is to argue that there are grounds for doubting that equal ontological commitment need be given to all parts of a successful scientific theory.

How are we to decide what entities we are committed to when we accept our best current scientific theories? Quine (1961, pp. 12-13) suggests when we accept a scientific theory we are committed to the sorts of things that would have to be in the range of values of variables of the theory's formulation in order for the theory to be true. 'To be is to be the value of a variable' as his famous slogan has it. But this suggests that the theory would first need to be formulated very formally. Perhaps not in the first-order logic that Quine favours, but in a logical language free of ambiguity, nonetheless. But scientific theories are seldom, if ever, formulated in this way. There is doubt that they could be. Much scientific talk invokes idealisations and is couched in the subjunctive mood. How such talk is to be formalised is highly controversial. The indispensability argument does not go through if it depends on our formalising scientific theories in order to determine their ontological commitments. Imagining what those commitments might be if the theories were to be formalised would be no better. Imaginary formalisations are a shaky foundation for platonism.

---

[5] Davis & Hersh attribute the percentages to J.D. Monk but I have been unable to track down the source. For an example of confused attempts by a highly reputable mathematician to sort out the ontological difficulties, see Mac Lane (1986).

At least some of the ontological commitments of scientific theories are quite clear. Current electron theory is committed to the existence of electrons. We know this from the history of the development of the theory. The investigations, theorising, and debate that followed the first postulation of electron theory were clearly aimed at discovering whether or not electrons exist. The success of crucial experiments established that they do exist. Physicists are clearly committed to the existence of electrons. If we accept their theories, then we too are committed to the existence of electrons.

This suggests that the way to discover the ontological commitments of scientific theories is to look to scientific practice.[6] This suggestion sits well with the naturalistic inclinations of those who put forward the indispensability argument. But when we examine actual scientific practice, we discover that it is in conflict with a crucial assumption of the indispensability argument. In order that the platonic claims be confirmed along with the non-platonic, all parts of a theory must be taken to be on an ontological and epistemic par.

However, scientists do not always interpret the different parts of a successful theory in the same way. As Maddy puts it:

> Historically, we find a wide range of attitudes toward the components of well-confirmed theories, from belief to grudging tolerance to outright rejection. (1992, p. 280)

She gives the example of atomic theory. Although it was well-confirmed by the mid-nineteenth century, it was not until the end of the century that scientists claimed to have verified the existence of atoms. As noted in Section 7.7, scientists only become seriously committed to the existence of postulated entities when they have successfully detected them, which involves causally interacting with them and, in particular, manipulating them. In the meantime, the scientists continued to use atomic theory as part of chemical theory because of its fruitfulness and explanatory power, and were prepared to believe the verifiable consequences of the theory. What some of them were not prepared to believe in was the existence of atoms even though that is clearly an indispensable part of atomic theory.

Perhaps those scientists who were sceptical of the existence of atoms were driven by positivist prejudice, having been overly influenced by the philosophy of their day. There is certainly evidence that this was the case for some of them. They subscribed to the doctrine that scientists need not, or should not, be committed to any of the 'unobservable' entities hypothesised by science. In fact, there was a wide range of opinion. Many others were

---

[6] The following argument draws on Maddy (1992, esp. pp. 280-82).

genuinely agnostic, and simply wanted more evidence. In the meantime, they found it convenient to regard atoms as 'chemical equivalents'—units of chemical reaction. Jean Perrin had provided what most regarded as conclusive evidence by 1913. Although Perrin was himself a believer, he does not seem to have regarded the initial doubters as irrational (Nye 1972).

The prevalence of positivist attitudes among nineteenth-century scientists raises a further problem for the postulationist position. Quine (1969) urges that 'first philosophy' should be eschewed when formulating epistemology.[7] We should look to the practice of current science to discover the norms of rationality for knowledge acquisition. But what are we to make of scientists who look to first philosophies like positivism or idealism to determine those norms? If Quine insists, in the face of such practice, that we, including the scientists, *ought* to accept the postulates of our current best science, then his own position smacks of first philosophy.

The development of atomic theory illustrates that, for practising scientists, indispensability alone does not establish truth, let alone provide grounds for knowledge. Scientists are not epistemological holists. If we are to look to science to guide our epistemology, then we should not be holists either.

When we look to the applications of mathematics to science, we find that mathematical entities are more likely to lie towards the sceptical end of the ontological-commitment spectrum. The bulk of these applications involve idealisations or approximations. For example, gases are analysed by assuming that their molecules are negligibly small with negligible inter-molecular forces. Surfaces are assumed to be frictionless. Fluids are assumed to consist of continuous matter and be infinitely deep. If such idealisations are not to be taken as literally true, why should the mathematics applied to them be so regarded?

Even more telling are mathematical entities that are 'invented' for a particular theory. For example, Newton proved that, for the purposes of his gravitational theory, a body may be treated as if all of its mass were at its centre of gravity. Consequently, Newtonians talk of 'point masses'. Point masses are mathematical constructs obtained from calculations in the integral calculus. They are indispensable to the theory. But Newtonians do not think that there are any point masses. They regard them as mathematical fictions.

---

[7] First philosophy is the *a priori*, armchair approach to philosophy, which sees philosophy as prior to empirically based knowledge.

Another example occurs in Ptolemy's cosmology. Early Ptolemaic systems featured an epicycle orbiting a deferent. Both epicycle and deferent were regarded as real existing spheres. Later, he introduced a system in which an epicycle orbited between an inner and an outer sphere. From this system of real spheres, the position of a deferent between the inner and outer spheres could be calculated. But the deferent of this system was a geometric fiction. Ptolemy did not suppose that it corresponded to the existence of yet another real sphere, although it was essential to his theory (Musgrave (1991, pp. 268 & 278, fn. 66).

Although the Newtonian and Ptolemaic theories are ontologically committed to these mathematical entities by Quine's criterion, the scientists themselves did not regard themselves as being committed to those commitments of their theories.

But perhaps these useful 'falsehoods' can only be understood in terms of more fundamental theories, which must be regarded as literally true. Isn't general relativity literally committed to the continuity of space-time, and committed in such a way that it is committed to the existence of real numbers? But are physicists committed to the continuity of space-time? Has this continuity been directly tested? It is doubtful that it could be. Perhaps the best reason for regarding space-time as continuous is that, if we do, the highly efficacious continuum mathematics can be applied to it. If scientists are agnostic about the continuity hypothesis, but adopt it because it is useful, then the mathematics that is applied to it must also be regarded as merely useful. The lack of direct verifiability is grounds for thinking that scientists are, and should be, agnostic.

Scientists are instrumentalists with respect to some elements of their theories, and realists with respect to others. But the distinction is not drawn on the sort of theoretical grounds beloved of philosophers of science. There is no fixed boundary between the observable and the unobservable, with realism on one side and anti-realism on the other (*pace* van Fraassen 1980). Nor is there one between entities and theory (*pace* Hacking 1983) and Cartwright 1983). Rather there are shifting and fuzzy boundaries between the unconfirmed and the confirmed, and between the merely confirmed and the verified, with the first boundary always ahead of the latter. Observable and unobservable entities, and existence and other theoretical claims, are to be found in all three categories. Postulations become better and better confirmed, and well-confirmed postulations become verified or known to be true when the right sort of interactions occur.

I have not demonstrated that scientists are, or should be, agnostic with respect to all the mathematics they use to formulate their theories. I have shown that ontological commitment does not follow from indispensability alone. More is required of the indispensability argument, if it is to establish a process for acquiring platonic knowledge. Given the variability of scientists' ontological commitment, the burden is now on the indispensabilist to demonstrate, by appeal to actual scientific practice, that scientists are seriously committed to the existence of platonic entities.

11.5. INDISPENSABILITY AND CAUSALITY

One reason for doubting that an indispensability argument for platonism can succeed is that in current science it is not clear what role mathematical entities play in even well-established scientific theories.[8] It is clear that the physical entities play a causal role, and this is particularly clear because scientists are explicit about that role. But scientists have little to say about mathematical entities. My personal observation is that they are at least as confused as mathematicians (see Section 11.3), although to be fair, some are quite explicit that they regard them as useful fictions. The prospect of understanding the role of useful fictions is promising. Field (1980, ch. 2) makes a start with examples such as applied arithmetic. Understanding the role of indispensable platonic entities is more problematic, especially from a naturalistic point of view.

Platonists have their reasons for supposing that there are entities that lack causal powers. But it is not clear that indispensabilists have good reason to suppose that mathematical entities are acausal. Scientists do not posit them as being acausal. In fact they are seldom, if ever, explicitly posited by scientists at all. They just appear in their theories. The properties of their explicitly posited objects are discussed and tested at length, but the metaphysical properties of mathematical entities scarcely warrant a mention. Even if their existence is indispensable, what grounds could there be for supposing that their acausality is indispensable? There are epistemological problems with acausal entities. Our epistemology may be insecure, but is it so insecure that our shaky metaphysics of mathematics should be allowed to cast doubt on it?

These considerations suggest that the Quine/Putnam indispensability argument cannot establish what its proponents intend. The form of the argument for this conclusion is simple. Suppose indispensability to science is the only good reason for believing in the existence of platonic objects. Either mathe-

---

[8] My arguments in this and the following section draw on Cheyne & Pigden (1996).

matical objects are dispensable to science, in which case there is no good reason for believing in the existence of platonic objects, or they are indispensable, in which case there is no good reason for believing that mathematical objects are genuinely platonic. Therefore, indispensability, whether true or false, does not support platonism.

As noted above, Hartry Field has set himself the task of demonstrating that mathematical objects are not indispensable to our best current science. There are two respects in which mathematical objects are supposed to be indispensable to science. They are indispensable when it comes to inference and they are indispensable in that our best scientific theories freely quantify over them. To prove that platonic objects are dispensable, and hence that we need not believe in them, Field has to do two things. First, he must show that platonic objects are not necessary for inference, and secondly, he must show that our best scientific theories can be nominalised, i.e. reformulated in such a way as to dispense with platonic entities.

I believe that Field has completed the first part of his project, although some regard this as a contentious claim.[9] He has demonstrated that if a nominalistic claim follows from a nominalised theory extended by a purely mathematical theory, then it follows from the nominalised theory alone. Thus, the mathematical theory need not be true in order that it be a useful aid to inference. But this result can only support the anti-platonist cause if our best scientific theories do not appeal to platonic objects. Hence the importance of the second part of Field's project. But this is incomplete. So far Field has produced a nominalised version of Newtonian gravitational mechanics. This theory is false and what is more has been shown to be so on empirical grounds. Complete and attractive nominalisations of General Relativity Theory and quantum mechanics (two current 'best' theories) are awaited.[10]

Two epistemic possibilities open up before us. The first is that Field's project will succeed and our best science will be nominalised. In that case the indispensability argument collapses and with it the best case for platonism. Alternatively, Field's project will fail and the best science will resist nominalisation. This is not to suppose that it will be demonstrated that nominalisation is logically impossible. (Indeed, true Quineans would be shocked at the idea.) Rather, as with the search for the Loch Ness monster, the persistent failure of our best efforts to find such a nominalisation will be sufficient reason to suppose that, like the Loch Ness monster, there is no such nominalisation to be found. In other words, we should conclude on

---

[9] For example, see Shapiro (1983) and Field's reply (1989, ch. 4).
[10] For recent attempts, see Mundy (1992) and Balaguer (1998b, ch. 6).

empirical grounds that theories that quantify over mathematical objects are better than their nominalistic rivals. In this scenario, Nature cries 'No!' to nominalised theories, but 'Maybe!' to ontologically loaded ones.

Now if our best science is condemned to quantify over mathematical objects, this would tend to show that such objects exist. But by the same token, platonism would seem to be in trouble. Why should theories that quantify over certain objects do better than theories that do not? One explanation is readily to hand. If we are genuinely unable to leave those objects out of our best theory of what the world is like (at least, that part of the world with which we causally interact), then they must be responsible in some way for that world's being the way it is. In other words, their indispensability is explained by the fact that they are causally affecting the world, however indirectly. The indispensability argument may yet be compelling, but it would seem to be a compelling argument for the existence of entities with causal powers.

Mark Colyvan (1998b) objects to this line of argument. He argues that this 'argument from indispensability to causal activity' lacks 'a crucial (and controversial) premise', namely, the assumption that 'all explanation is causal explanation' (p. 116). Colyvan's objection overlooks the precise thrust of the argument. The conclusion of the argument is that the only plausible explanation for the indispensability of mathematical objects is (as far as I can see) a causal explanation. This leaves platonists faced with the challenge to produce a plausible, non-causal explanation for the indispensability of mathematical objects. My argument does not require the assumption that all explanations are causal. Scientific theories explain causal facts. I am suggesting that if certain entities are indispensable to scientific theories, then this argues for the causality of those entities.

Why couldn't a mathematical object be a constituent of a causal fact (or event or state) and yet itself be causally inert? Perhaps it could. But either its presence would make no difference to the effects of that fact and so any mention of it could be omitted from an explanation of those effects, or its presence would make a difference to the effects of the fact in which case it would be perverse to deny it causal efficacy. For example, suppose the fact that there are three cigarette butts in the ashtray causes Sherlock to deduce that Moriarty is the murderer, and that if there had been more or fewer butts he would have deduced otherwise. The fact that there are three cigarette butts in the ashtray is clearly causal. Suppose that the number three is an indispensable constituent of that fact.[11] Could platonists then claim that the number three is an acausal constituent of the fact? On the face of it, no. It's being a constituent of the fact makes a causal difference. If the number two or the number four were in its place, the effects would differ. What more is needed for it to qualify as an object with causal powers?

Colyvan objects to this argument on the grounds that it 'seems to revolve around a fairly undiscriminating notion of causation' (1998b, p. 116). The problem with such a notion, he believes, is that:

> [i]t implies, for instance, that the angle sum of a triangle *causes* bodies to be accelerated, since if the angle sum of a triangle is π radians, the space is (locally) Euclidean and so massive bodies experience no net force, if the angle sum is not π radians the space would be non-Euclidean and hence any massive body would be experiencing a net force. Thus, if there were a change in the angle sum of a triangle, the future light cone of the world would be different, in that it would contain an accelerated body. (p. 117)

This example is, in essence, the same as my Sherlock and Moriarty example, so my response must be the same. I do not appeal to a 'simple counterfactual dependence theory of causation' as Colyvan suggests (p.117). Suppose space is non-Euclidean and that this fact plays a role in a causal explanation of the acceleration of a certain massive body. Suppose that a particular triangle in that space has an angle sum of $(\pi + \partial)$ radians. If that triangle had a different angle sum, then (given certain assumptions) the body's acceleration would have been different. I agree that it does not follow from the truth of such a counterfactual alone that the angle sum causes the body to accelerate. But suppose (wildly improbable though it is) that the number $(\pi + \partial)$ is an entity whose existence is indispensable to our best explanation of the body's acceleration. It does follow that if this supposition were correct then platonists would face the challenge of explaining how it is that this entity is both indispensable and acausal. Colyvan overlooks the crucial role of indispensability in my argument.

11.6. A CHALLENGE TO PLATONISTS

The challenge to platonists is for them to provide an explanation for the indispensability of objects whose presence (they claim) makes no causal

---

[11] It isn't, of course. Frege has shown how we can say that there are three butts in the ashtray without reference to the number three. But if we want an example in which indispensability is more likely, we shall need to delve into the realms of General Relativity or quantum mechanics. If platonists believe that they can strengthen their case with such an example, I look forward to seeing it.

difference. And it will need to be a better explanation than the suggestion that they are indispensable because their presence does make a causal difference.

So whether or not Field's project succeeds, platonism seems to be in trouble. Either the project succeeds and the indispensability argument must be abandoned, or the project fails and, although there is good reason to believe in mathematical objects, there is also good reason to believe that they are not acausal. Either way, platonism faces a challenge.

The challenge is most squarely directed at those platonists who believe that the indispensability argument provides our best reason for adopting platonism. Does this mean that other platonists ('dispensabilist platonists' perhaps) are off the hook? Well, it will still be a challenge (if not such a serious one) to those who believe that indispensability provides some reason for believing platonism. And some platonists do appear to believe this, otherwise they would not take the trouble to argue that Field's project cannot succeed.[12]

What about those who claim that the indispensability argument gives us no reason for believing in platonism? Presumably they will have some other reason for believing in the existence of platonic objects. Their grounds for adopting platonism will be either empirical or *a priori*. If they are empirical, then a similar argument may be run against them, and they would face a similar challenge. If *a priori*, then they should not be surprised if mathematics turns out to be dispensable. On the other hand, if it proves to be indispensable, they face the challenge of providing an explanation of this fact, but one that should not appeal, however indirectly, to the causal efficacy of mathematical objects.

Many platonists believe that the success of Field's project would count against platonism. And in this they are surely correct. But what they do not realise is that the failure of Field's project would also count against platonism. For it would create a problem that platonists are ill-equipped to solve—how to account for the indispensability of numbers in describing the causal nexus whilst absolving those numbers from the sordid taint of causality. One option is what might be called the neo-Kantian or 'framework' solution. Numbers are needed to underwrite any conceivable causal order but they themselves play no part in the proceedings. They provide a sort of metaphysical framework for any possible physics—an indispensable, indeed, a necessary backdrop for the causal show. But though there could be no

---

[12] See for example Hale (1987 & 1990), Hale & Wright (1992), Resnik (1990) and Shapiro (1983).

causal structure without numbers, numbers are not implicated in the causal shenanigans described by any science whether actual or merely possible.

This theory requires a lot of work if it is to be anything more than a collection of figures of speech. Imagine a conversation between a neo-Kantian and a framework-sceptic. 'How can numbers play a necessary part in causal explanations even though they exercise no causal powers?' 'Well, they're part of the framework.' 'What is this framework?' 'Well, of course, "framework" is only a metaphor, since in the real world frameworks actually do a lot of causal work, but what I mean by "framework" is a kind of a thing which ... well, um, ... what it does is it allows numbers to play a necessary part in causal explanations even though they exercise no causal powers.' 'Gee, thanks!'[13]

Insofar as sense can be made of the framework theory, it is false. Field's achievements as a nominaliser have demonstrated this. For he has succeeded in nominalising Newtonian physics. This physics describes a simpler set of worlds than the one we actually inhabit. In these Newtonian worlds there are causal laws and causal histories but numbers are superfluous to requirements. They are not needed to underpin the causal goings-on, and if they exist at all, they constitute an infinity of spare parts, of underpinnings that underpin nothing. Far from constituting a necessary framework for any conceivable physics, they turn out to be unnecessary to the physics that everyone believed in until Einstein came along. According to the framework theory, numbers are necessary because they are presupposed by any conceivable causal system. By nominalising Newtonian physics, Field shows that this is not so. For we can conceive of a causal order (namely that described by Newton) which can do without the framework.

The new problem that the platonists face is this: How can a set of necessary beings help explain a contingent set of facts (namely the facts accounted for by Einsteinian physics) when they would not be needed if the facts were otherwise (i.e. such as to confirm Newtonian physics)? Numbers would be like a modally capricious God who in some worlds stoops to create whilst in others he prefers to reign in splendid isolation. Such a God might exist necessarily, but his relational properties would be contingent. For in some worlds he would be causally active and in others not. So too with numbers. If they are needed to account for the goings on in some worlds and not others, this suggests that they are causally active in some worlds and not others. And if Field's project fails, this suggests that one of the worlds in which numbers are causally active is the actual one. Given the success of

---

[13] Cf. Stove (1991, p. 53) on the synthetic *a priori*.

Field's project so far, the ultimate failure of his enterprise would be just as damaging to platonism as his total triumph.

Colyvan objects that although this point 'is very telling on any platonist that holds the "necessary framework" view...it completely misses the target of the Quinean indispensabilist who denies that mathematical entities are necessary' (1998b, p. 118). Fair enough. But then the challenge becomes: 'How could a causally-inert contingent framework play an indispensable role in our best scientific theories?' Colyvan does not suggest an answer and clearly admits as much. He is content to assert that such a framework has 'the makings of a very good reply to the challenge' (p. 118). But with nothing said to back this claim, it has a distinct air of whistling in the wind.

Recently, Ruth Richardson, a former finance minister in New Zealand, published a memoir under the title *Making a Difference*. Her implied boast was that she (unlike most finance ministers) was causally efficacious. The indispensability argument claims that numbers, sets, etc. make a difference (which is why they cannot be dispensed with). But it is difficult to see how they can do this without being causally efficacious. Hence numbers, if they are to be believable, must be like Ruth Richardson.

To sum up the state of play with respect to indispensabilist arguments, the role of mathematical entities in scientific discourse is uncertain. There are doubts as to their indispensability. There are doubts as to the degree of ontological commitment accorded them. There are doubts as to their acausality. So long as the role of mathematical entities in scientific discourse is so uncertain, the claim that we acquire platonic knowledge via scientific practice cannot be made out. The burden of proof still lies firmly on the platonists.

CHAPTER 12

PROBLEMS WITH PROLIFIGATE PLATONISM

12.1. FULL-BLOODED PLATONISM

Variations on a new platonist epistemology have been offered recently. Mark Balaguer (1995, 1998a & 1998b) offers one version, and Bernard Linsky and Edward Zalta (1995) another. Although there are important differences between their proposals, what they have in common is the suggestion that if there is a plenitude of mathematical entities, then there is no problem about acquiring knowledge of them. I first discuss Balaguer's proposal and then discuss the extent to which Linsky and Zalta's account faces similar problems.[1]

Recall Benacerraf's challenge to platonism. If mathematical objects are acausal and if a naturalistic account of human knowers is true, then we cannot have knowledge of the existence and nature of those objects. There is a tension between the acausality of platonic entities and a naturalistic epistemology. Platonists owe us an account of how we acquire knowledge of the mathematical realm that they claim exists.

Balaguer presents his account as a direct response to Benacerraf's challenge. Suppose that there is a plenitude of mathematical objects, in the sense that all mathematical objects that could possibly exist actually do exist. Suppose, in other words, that what Balaguer calls *full-blooded platonism* (FBP) is true. In that case, he argues, any consistent purely mathematical theory truly describes some part of the mathematical realm. It follows that we can attain knowledge of any mathematical object simply by thinking consistently about it. Since mathematicians can avoid inconsistency, there is

---

[1] My discussion of Balaguer's proposal is based on Cheyne (1999).

no problem about their acquiring reliable mathematical beliefs and, hence, mathematical knowledge. If FBP is true, then Benacerraf's challenge is met. According to Balaguer, this conditional is all that is required to save mathematical platonism from Benacerraf's challenge. He does not need to claim or establish that FBP itself is true and does not seek to do so.

I have two objections to Balaguer's position. The first begins by noting that Benacerraf's challenge is two-pronged. It has a semantic component as well as an epistemological one. For a purely mathematical theory to successfully describe some part of the mathematical realm, the terms of that theory must successfully refer to objects in the mathematical realm. But it is unclear how human language could refer to such abstract objects. Balaguer's platonist epistemology depends crucially on successful reference. Without an adequate response to the problem of reference for abstract objects his epistemology collapses. I argue that no such response is available.

My second objection is that, even if Balaguer's conditional could be established, this would not be sufficient to save mathematical platonism from all epistemological problems. There is no way we can know that a platonic realm, plenitudinous or not, actually exists. So we still have no reason to accept mathematical platonism.

First I outline in more detail Balaguer's epistemology for FBP and his account of reference. I show why a satisfactory account of reference for FBP is crucial for a successful response to Benacerraf and demonstrate that Balaguer's account is unsatisfactory. Next I suggest that a less restricted view of successful reference in a mathematical context may provide a more successful account. Unfortunately, the result is an 'anaemic' platonism that is not robust enough to qualify as a genuine platonism and, thus, cannot meet Benacerraf's challenge. Finally I advance my second objection to Balaguer's proposal, before discussing the proposal of Linsky and Zalta.

12.2. THE EPISTEMOLOGY AND SEMANTICS OF FBP

Because he considers the concept of knowledge to be imprecise and controversial, Hartry Field (1989, pp. 230-31) suggests that the challenge to platonists is better expressed as the challenge for them to account for the supposed fact that (as a general rule)

(MA)   If mathematicians accept that p, then p,

where '*p*' is a mathematical statement. Balaguer accepts the challenge in these terms. He sets out his argument that FBP-ists can meet this challenge as follows:

(i)   FBP-ists can account for the fact that human beings can (without coming into contact with the mathematical realm) formulate purely mathematical theories.

(ii)   FBP-ists can account for the fact that human beings can (without coming into contact with the mathematical realm) know of many of these purely mathematical theories that they are consistent.

(iii)   If (ii) is true, then FBP-ists can account for the fact that (as a general rule) if mathematicians accept a purely mathematical theory T, then T is consistent.

Therefore,

(iv)   FBP-ists can account for the fact that (as a general rule) if mathematicians accept a purely mathematical theory T, then T is consistent.

(v)   If FBP is true, then any consistent purely mathematical theory truly describes part of the mathematical realm, that is, truly describes some collection of mathematical objects.

Therefore,

(vi)   FBP-ists can account for the fact that (as a general rule) if mathematicians accept a purely mathematical sentence *p*, then *p* truly describes part of the mathematical realm. (Balaguer 1998b, pp. 51-52)

Balaguer claims that premise (i) is trivial. It is analogous to the claim that human beings can formulate fictions and myths without coming into contact with fictional or mythical entities. Regarding (ii), Balaguer says that platonists can simply give any account of our knowledge of mathematical consistency that anti-platonists can. He points out that we do not need any access to a set of objects (mathematical or physical) in order to know whether a set of sentences about those objects is consistent. He claims that (iii) is fairly trivial. Apart from certain qualifications, I shall not take issue with (i), (ii) and (iii), nor with Balaguer's claim that (vi) follows fairly trivially from (iv) and (v). For the most part, it is premise (v) with which I take issue.

Here is a sketch of Balaguer's semantics for natural-number arithmetic.[2] I start with the following formulation of the Peano postulates that purport to describe the natural numbers:

(P1)   Zero is a natural number.

(P2)   Every natural number has a successor, which is also a natural number.

(P3)   Zero is not the successor of any natural number.

(P4)   Different natural numbers have different successors.

(P5)   [The induction postulate]

Initially, suppose this language to be uninterpreted. Then suppose that all the non-mathematical terms are assigned their usual meanings in ordinary English. The only mathematical terms are 'zero', 'natural number', and 'successor'. Standard syntax for ordinary English tells us that 'zero' is a singular term, 'natural number' is a one-place predicate, and 'successor' is a one-argument functor. (If it doesn't tell us that, then we can stipulate them as such.) According to standard semantics, objects are assigned to singular terms, sets of objects to one-place predicates, and sets of ordered pairs of objects to one-argument functors. Given (P1-5) and a domain of infinitely-many objects, there are infinitely-many ways of carrying out such an assignment (but no ways for a finite domain). Each assignment will pick out an ω-sequence of objects or, perhaps more accurately, each assignment gives rise to an ω-sequence.

If the domain is the abstract realm of FBP, then, clearly, we must not think of these assignments as human operations. Our isolation from the realm ensures that we cannot carry out such assignments. Rather, we must assume that these assignments exist independently of us and of our activities. If no further restrictions are placed on these assignments, then any mathematical object could be assigned to 'zero', and any other object could be assigned as its successor, and any object other than those two could be assigned as the successor of the successor of the object assigned to 'zero', and so on. Thus, given no further interpretation of the Peano postulates, any

---

[2] This sketch should be seen as a rational reconstruction, gleaned from his (1995), (1998a) & (1998b). He may not agree with all details.

mathematical object could occupy any position in infinitely-many ω-sequences.

However, according to Balaguer, there are further restrictions. There is more to our notion of the natural numbers than satisfaction of (P1-5) with the mathematical terms taken as uninterpreted primitives. We have what he calls our *full conception of the natural numbers* (FCNN). FCNN contains various notions that we share as a community as to what sort of things natural numbers could, or could not, be. In particular, natural numbers are abstract, they have no physical properties, but they differ from other mathematical objects such as sets and functions. This narrows the field. Any ω-sequence containing an object that does not satisfy FCNN is ruled out. Furthermore, FCNN may also include a notion as to what sort of relation qualifies as a successor relation, so that any sequence in which the objects are not genuine successors of their predecessors would be ruled out.

However that may be, no matter how much FCNN narrows the field, it lacks the resources to pick out a unique ω-sequence. There are properties which mathematical objects may possess, but which we have not, and never will have, ever considered. But, given the profligacy of FBP, for any conception we may have of a mathematical object, there will exist (infinitely-many) objects which satisfy our conception but which differ among themselves with respect to unconsidered properties. Such properties will not be part of FCNN, so FCNN will not distinguish between objects with those properties and objects without them. In other words, it appears that we lack the resources to uniquely specify individual objects in a full-blooded platonic realm. But mathematical talk is replete with singular terms that, on the face of it, are intended to refer to unique objects.

Balaguer discusses this problem and offers a solution. Briefly, he argues that while mathematical theories 'truly describe collections of abstract objects, they do not pick out *unique* collections of such objects' (1998b, p. 73), and that such non-unique reference is not a problem in this (the mathematical) context. He calls this position *non-unique platonism* (NUP) and argues that all platonists (not just FBP-ists) should embrace it. Applying NUP to FCNN, each assignment of the terms will pick out an ω-sequence of natural numbers, including one and only one zero. Within the context of any assignment, the singular term 'zero' will refer to a unique object. However, there will be infinitely-many, equally-good assignments in which 'zero' refers to another object. It is in this wider context that 'zero' refers non-uniquely. But, according to Balaguer, this non-unique reference is benign. Whatever the theory says under one assignment, is equally the case under all. For the most part, it does not matter that we talk as though there is only one unique set of natural numbers. Such talk is a convenient and harmless fiction.[3]

So much for reference. As far as the truth-values of arithmetic statements are concerned, Balaguer seems to have in mind a variety of supervaluation. To illustrate this, it will be convenient to extend the theory of Peano arithmetic by adding the usual postulates that define addition and multi-plication (in terms of zero and the successor function). We also add the usual vocabulary consisting of the numerals and terms such as 'odd', 'even', 'prime', 'square of', etc. (defined in terms of previously defined terms). FCNN picks out certain ω-sequences. For any arithmetic sentence, either the sentence will be true of all those ω-sequences (e.g. '17 is prime'), or it will be false of all of them (e.g. 'the sum of 3 and 4 is 9'), or it will be true of some and false of others (e.g. an attribution of a 'never-considered' property to a natural number). Those sentences that are true of all the ω-sequences are true, those that are false of all the ω-sequences are false, and those that are true of some and false of others are neither true nor false.[4]

Balaguer maintains that it is plausible to suppose that the arithmetic of FCNN is categorical, from which it follows that those sentences that lack a truth-value are unproblematic because they lack arithmetic interest, they are not relevant in an arithmetic context. Things are not so straightforward in the case of set theory and the continuum hypothesis (CH) since he wishes to leave open the possibility that our full conception of sets is not categorical. So, by the above account, CH would appear to lack a truth value in standard set theory. Yet it is of considerable set-theoretical interest. Balaguer discusses this in some detail (1998b, pp. 62-64) and I return to this issue at the end of the next section.

My purpose in this section has been to sketch a neutrally account of Balaguer's epistemology and semantics for his full-blooded platonism.

12.3. PROBLEMS WITH REFERENCE FOR FBP

Before turning to my chief objection to Balaguer's epistemology for FBP, I note an apparent internal inconsistency for his theory. According to

---

[3] There is an analogy here with modal-realist talk. We may speak loosely of *the* possible world in which the Spice Girls sing Mozart opera, although there are infinitely many such worlds.

[4] For detailed accounts of supervaluational semantics, see van Fraassen (1969) and Fine (1975).

Balaguer, given FBP, all consistent mathematical theories are true, because all consistent mathematical theories describe part of the abstract mathematical realm. But suppose we have a mathematical theory that insists that the natural numbers (or other objects rather like the natural numbers) are unique. In other words, suppose it is part of our full conception of certain mathematical objects that they are unique. Such a theory would be consistent. But then, given FBP, this theory would be false. It would not describe part of the mathematical realm because no conception can capture unique objects. So not all consistent mathematical theories are true.

Balaguer's response is to argue that the notion that a mathematical theory is, or could be, about a unique set of objects is misguided or naïve (1998b, pp. 88-89). Even given a more traditional platonism, it is unreasonable to suppose that our mathematical theories would be satisfied uniquely in the platonic realm. Besides, any differences between the different objects that satisfy the theory would not be relevant in a mathematical context. He points out that mathematical practice tells us that if it were pointed out to mathematicians that their theories cannot refer uniquely because of properties we have never imagined, then they would not think that mattered in the least. Uniqueness may be part of an 'untutored' conception of the natural numbers, but it could not be part of our 'educated' conception of the natural numbers.

One way he might state his position is to claim that no theory that insisted on strict uniqueness could be a genuinely *mathematical* theory. In that way, he can maintain his claim that, given FBP, all consistent mathematical theories are true. Fair enough. But to rule out certain notions on the grounds that they are 'untutored' can make one's position a hostage to fortune. It leaves open the possibility that further 'tutoring' may rule one's own position out of order. I return to this possibility in Section 12.4.

My chief objection to Balaguer's epistemology for FBP is that its account of reference cannot avoid Benacerraf's problem of reference for mathematical platonism. According to Balaguer, '[i]n giving a standard semantics for the language of arithmetic, what we do is assign objects to the singular terms, sets of objects to the one-place predicates, sets of ordered pairs of objects to the two-place predicates, and so on.' (1998b, p. 89). Now such assignments can be made in two ways – what I shall call 'intensionally' and 'extensionally'. An example of an extensional assignment would be if I were to choose an arbitrary subset of sheep from a mob and declare them, and only them, to be 'slithy' sheep. Intensional assignment occurs when I use the word 'merino' to pick out the subset of merino sheep from the mob. Roughly speaking, in the first case the chosen set gives meaning (or, at least, reference) to the term, and in the second case, the meaning of the term chooses (or enables reference to) the set. Some cases have both intensional and extensional aspects. For example, I may define the term 'mimsy' as referring to all and only the two-toothed, merino ewes in the mob. Here, a meaningful expression picks out a set, which is then assigned to a previously meaningless term. For my purposes, the important distinction will be between those cases that are purely extensional and those that involve at least some intensional aspect, so I shall extend the term 'intensional' to include the latter.

The above quotation from Balaguer has more than a whiff of the purely extensional about it. However, it is clear from the rest of his writings that he intends that reference for FBP is intensionally acquired. In particular, for basic terms like 'number' and 'set' we have a standard conception as to what sort of things numbers and sets are, and if we use the terms standardly, then we successfully refer to numbers and sets in the abstract realm. But how can this be? I can successfully refer to the merino sheep in Iran, without relying on any contact with the sheep of that remote country. This is because of my contact with other merino sheep or with someone who has had contact with merino sheep. But if numbers are abstract objects then no-one has had contact with them. My objection does not depend on a narrow causal theory of reference. I am willing to concede that we may successfully refer to unicorns, should they exist, even though no-one has ever had contact with them.[5] According to a broader (causal-descriptive) theory, we can have a conception of unicorns in terms of the familiar properties of being a horse and having a single horn.

What are the number-like properties supposedly possessed by numbers? First of all, they are abstract objects. Now, if by 'abstract object' we mean an object that is non-physical, non-spatiotemporal and acausal, then the term is meaningful and if such objects exist, then we can successfully refer to them. But beyond that it is difficult to see how we might use language to successfully discriminate between such objects, since that discrimination must be on the basis of non-physical properties and relations. The situation is analogous to our inability to discriminate on the basis of physical properties that we are unaware of. If the human race had never developed sight, then our language could not have included meaningful colour terms, unless an alternative means of accessing colours had become available.

---

[5] Narrow causal theorists claim that if unicorn-like creatures were discovered, they would not be the referents of our word 'unicorn'.

Balaguer suggests that part of our conception of numbers is that they are not sets or functions, etc. This, in itself, does not help to fix the reference of 'number', unless we can already successfully refer to sets and functions, etc. To say that sets are not numbers or functions or whatever, and functions are not numbers or sets or whatever, and so on, is merely circular. We need some way of breaking into the circle.

Insofar as he is aware of this problem, Balaguer suggests that FBP-ists can proceed by appealing to some of the familiar notions we apply to the physical (non-abstract) world. Perhaps our counting practices and our acquaintance with ordered lines of succession and of collections of concrete objects will suffice to provide us with concepts that characterise numbers. But our counting practices and notions of ordered succession are heavily reliant on spatiotemporal and causal relations. There are, in the physical world, what we may call 'natural quasi-ω-sequences'. For example, a mob of sheep may be ordered by weight or time of birth or the order in which they leave the shearing shed. And such sequences, if not actually infinite, can continue indefinitely. A mob of sheep may continue through successive generations.

Maybe there are analogous 'natural' sequences in the profligate realm posited by FBP; sequences based on non-physical, rather than physical, relations. But given the profligacy of that realm, we should expect them to be even more ubiquitous than in the physical world. What will distinguish the sequences of natural numbers from all the rest? In the mob-of-sheep example, note that the same sheep could occupy the '3 position' of one sequence and the '4 position' of another. Could not the same thing happen in the abstract realm, resulting in some object being both 3 and 4, and hence, both odd and even? Not so, according to Balaguer. An object could occupy both the '3 position' and the '4 position' of different sequences, but only one of those sequences could be a sequence of natural numbers. This is because in order to satisfy FCNN, an object occupying the '3 position' of a sequence of natural numbers must have the property of being 3, and any object with that property cannot have the property of being 4. This highlights, once again, the problem of reference for FBP. How does our language, of itself, refer to objects which possess the property of being 3? We can insist that the slithy sheep in a remote mob cannot also be mimsy, because it is part of our conception of slithy sheep that they have the property of being slithy, but do not have the property of being mimsy. It does not follow that there is any fact of the matter as to which sheep are slithy or as to which of the sheep the word 'slithy' refers.[6]

FBP-ists may seem to have an easier task convincing us that our language can refer successfully to abstract sets. Balaguer (1998b, p. 87) claims that ø and {ø} are clearly distinguishable on the grounds that it is part of the nature of the latter that it contains the former. Presumably it is also part of the nature of ø that it is empty. But in the physical realm, the notions of emptiness and containment cover a wide range of context-dependent alternatives. Given any two objects, the question of whether one contains the other will often depend on exactly which relation of containment we have in mind. Similarly, for the property of being empty. The profligacy of the abstract realm of FBP and the fact that the properties and relations in question are not *physical* emptiness and containment, but merely abstract analogues, must give rise to even greater indeterminacy. Even if our notions of emptiness and containment can rule out some of the abstract objects as inappropriate referents of 'ø' and '{ø}' (and I don't believe that they can), there is no reason to disbelieve that there are pairs of objects to which the terms may refer interchangeably. Given our isolation from the realm, there is no way of resolving such ambiguities as we can, and do, in the physical realm.

To sum up, given the existence of the abstract realm of FBP in all its variegated glory, as far as we and our linguistic practices are concerned its occupants can be little more than 'bare particulars', since our isolation from them means that we lack the resources to discriminate between them in any interesting way, and certainly not in the detail required by successful mathematical talk.[7]

Recall Balaguer's crucial premise (v):

(v)   If FBP is true, then any consistent purely mathematical theory truly describes part of the mathematical realm.

The thrust of my argument is that mathematical talk can no more truly describe part of the mathematical realm of FBP than Lewis Carroll's *Jabberwocky* can describe part of David Lewis's realm of possible worlds. Unless the reference of the terms are fixed in some way, there can be no fact

---

[6] I assume that 'slithy' and 'mimsy' are nonsense words and disregard Humpty-Dumpty's gloss on them.

[7] I am not saying that the objects in question *are* bare particulars. (How, in consistency, could I?) Rather, I claim that for all we can say or know, they might as well be. See Balaguer (1998b, p. 87).

of the matter as to whether or not any possible world is inhabited by the Jabberwock or by slithy toves.

FBP-ists may respond to my argument by providing an account of how we can successfully refer to the denizens of their populous realm or by arguing that they are not under an obligation to do so. Balaguer has not yet provided the former and I next argue that the latter course is not open to him. As already noted, Benacerraf's challenge to platonism is two-pronged and the semantic prong closely parallels the epistemic one. Given our causal isolation from any platonic realm, it is problematic how we can refer to, or have knowledge of, platonic objects. The burden of proof is on platonists to explain how such reference and knowledge is achievable. Some platonists reject the challenge on the grounds that it is premised on suspect theories of reference and/or knowledge. They argue that the burden of proof is on anti-platonists to articulate and establish those theories. They claim that our ability to discuss and acquire knowledge of mathematics is already more firmly established than any extant epistemology or theory of reference, all of which are hotly contested.[8]

Balaguer accepts the burden of providing an epistemology for platonism. But in so doing, he cannot shrug off the burden of providing a semantics for his platonism. He cannot argue that the problem of reference is also a problem for any (standard) platonism, and avoid the problem on the grounds that he is not duty bound to provide solutions to all the problems facing platonism. The problems facing platonistic epistemology and semantics are too closely intertwined to allow that as an option.

In the next section I discuss a possible 'way-out' for profligate platonism, but first I return to a problem mentioned above concerning set theory and the continuum hypothesis (CH). Suppose (a big 'suppose' in view of what I have just argued) FBP-ists can provide a convincing account of how we can successfully restrict the reference of our mathematical terms to the 'right' sort of objects, albeit that the singular terms will refer non-uniquely. Now, if a singular term such as 'the number 17' refers non-uniquely, there appears to be a threat of contradiction. If the ω-sequences picked out by FCNN differ from each other, then the different number 17s will have differing properties and bear differing relations to other members of their sequence. But if the

---

[8] Burgess and Rosen (1997, Part I.A.2) survey this debate in detail and conclude that 'in so many ways [debate over reference] is just a replay of debate over knowledge' (p. 49), and that there is a 'tendency [for] nominalist arguments and anti-nominalist counter-arguments to reach stalemate over burden of proof' (p. 60).

singular term 'the number 17' denotes all of those 17s, and not just one of them, then there are properties P such that:

(E)   The number 17 is P and the number 17 is not P,

so we have an apparent contradiction. Recall that Balaguer's response is that these contradictory properties will all be properties we have never thought of, or that are not of mathematical interest or importance, or are, at least, not relevant in the context of standard, natural-number arithmetic. Thus, given a supervaluation semantics, each of the conjuncts of (E) will lack a truth-value and so will not contradict each other, while (E) itself will be false.

The hope is that contradiction can be avoided by our stipulating what is and isn't relevant in a particular mathematical context. But consider set theory and the continuum hypothesis. There was a time when CH had not been thought of. Then there was a period when it was entertained but its status was unclear. During that period, any standard interpretation of the language of sets (for example, Zermelo-Fraenkel) would continue to pick out collections of sets of which CH would be true for some but not others. Now that we know that CH is independent of the ZF axioms, perhaps we can decide whether or not to include it in our new standard interpretation. But there was a time when, given that the term 'the empty set' referred non-uniquely, it was the case that the empty set was and was not the first term of a sequence of sets for which CH held.

Although this may not be an insoluble problem for FBP-ists, it does raise the worry that a supervaluational semantics in a mathematical context may not yield a notion of truth which is quite as ordinary as Balaguer insists (1995, p. 315) or that 'fits perfectly' with the way mathematicians use the word 'true' (1998b, pp. 59-60). I shall not say more about this worry. My main case against FBP is the lack of an adequate account of reference in a platonistic context.

12.4. ANAEMIC PLATONISM

I have argued that, given our isolation from the platonic realm, we ought to be sceptical that the singular terms and predicates of our mathematical language can successfully refer to specific objects and properties in that realm. If we grant the viability of non-unique reference, then we have to allow that every mathematical singular term refers (non-uniquely) to every mathematical object and every mathematical predicate applies (non-uniquely) to every set of mathematical objects (or of ordered pairs of objects, etc.). Balaguer appears to agree that 'this *vicious* sort of non-unique reference is surely unacceptable' (1998b, p. 87) and defends his position against the charge. But why is it unacceptable? I suggest that if non-uniqueness really is a non-problem for platonism, then non-uniqueness on this more widespread level should also be a non-problem.

In fact, such 'vicious' or unrestricted non-unique reference actually follows from FBP, as I now demonstrate. According to FBP, all consistent mathematical theories are true. I shall state a consistent mathematical theory called *djinn arithmetic*.[9] Djinn arithmetic, like natural number arithmetic, has five axioms, but there is more to its full conception than that. However, its full conception is considerably more streamlined than FCNN and this is what makes unrestricted reference unavoidable (even if more restricted reference were possible). The axioms are:

(D1)   Genie is a djinn.

(D2)   Every djinn has a meta, which is also a djinn.

(D3)   Genie is not the meta of any djinn.

(D4)   Different djinns have different metas.

(D5)   [A suitable induction postulate]

In addition, according to the full conception, all djinns are mathematical objects distinct from other kinds of mathematical objects,[10] but otherwise the terms 'Genie', 'djinn', and 'meta' are free of any connotations. Note that it is part of the full conception that the terms are connotation-free. Taking as the domain the abstract realm of FBP, if the terms are assigned referents in the standard way, and a supervaluational semantics applied such that any statement in djinn arithmetic is true if it is true for every assignment, and false if false for every assignment, and lacks a truth value otherwise, then we have a consistent, true mathematical theory with a semantics according to which any mathematical object can play the role of any object in djinn arithmetic. So unrestricted reference had better be acceptable to FBP-ists, as it follows from the tenets of FBP.

It might be objected that djinn arithmetic is not a genuine mathematical theory, on the grounds that mathematicians neither intend nor would they

---

[9] From Hofstadter (1979), p. 216.
[10] This ensures that we have a mathematical theory on our hands and that the theory can be combined with other mathematical theories.

allow 'vicious' reference, or on some other grounds. But note that djinn arithmetic is isomorphic with Peano arithmetic. For every theorem of djinn arithmetic there is an equivalent theorem of Peano arithmetic and vice versa. If we give up the notion that FCNN does any useful work (as I have argued is the case in an FBP context), then we can argue that djinn arithmetic just *is* natural number arithmetic. FCNN becomes part of an 'untutored' conception of the natural numbers. The fact that ω-sequences can overlap so that one and the same object can be a particular djinn/natural number on some assignments and a different djinn/natural number on others is just not relevant in a mathematical context. If non-unique reference is unproblematic, then unrestricted non-unique reference is likewise unproblematic.

Let us explore FBP with unrestricted reference a little further. Notice that the myriad properties and relations supposedly instantiated by the objects in the abstract realm play no part in the truth of arithmetic statements. Similarly for the theorems of any other consistent mathematical theory. That being so, the full-bloodedness of FBP becomes otiose. A domain of numerically distinct 'bare particulars' will do the job just as well, as long as there are enough of them. If Benacerraf's challenge can be met, then it can be met by what we may call *anaemic platonism* or AP. AP is (roughly) the view that as many mathematical objects as could possibly exist actually do exist.

Although *prima facie*, anaemic platonism meets Benacerraf's challenge, it is fairly obvious why it is unsatisfactory. AP is not really a platonism at all or, at least, it is not a *mathematical* platonism. As far as AP is concerned, a mathematical object might as well be any object that can play the role of a mathematical object, since any properties or relations it has are irrelevant to that role. It follows that any object (abstract or concrete) is (potentially) a mathematical object. Mathematical platonism includes the claim that mathematical knowledge is about mathematical objects, and this claim is hopelessly vitiated if every object is a mathematical object. Furthermore, mathematical truth, according to AP, amounts to no more than that of having a model. If semantic consistency is construed as having a model, then AP-truth is equivalent to semantic consistency and the notion of mathematical truth for AP is not the robust or standard notion one expects of a mathematical platonism.[11]

---

[11] I observe that if we can construe semantic consistency non-platonistically as the *de dicto* possibility of having a model, then it is an (apparently) small step from an AP account of mathematical truth to a modal account, whereby a mathematical theory is true just in case it is possible that it has a model.

I conclude that in the absence of an adequate account of reference, full-blooded platonism cannot provide an epistemology that meets Benacerraf's challenge. We do not have the resources to restrict the reference of our terms when applied to a realm of acausal, platonic objects. If we allow unrestricted reference, the result is too anaemic to qualify as a genuine mathematical platonism.

12.5. TRUE BELIEF WITHOUT KNOWLEDGE

Suppose these semantic issues I raise were to be satisfactorily resolved. Recall that Balaguer's task was to meet Field's challenge to account for the fact that (as a general rule)

(MA)   If mathematicians accept that p, then p,

where 'p' is a mathematical statement.

Let us suppose that Balaguer has indeed met that challenge. Would he then have succeeded in putting the epistemological issue to rest? I think not. The motivation for Field's challenge is that he doesn't wish his objection to platonism to become bogged down in a dispute over the concept of knowledge. So he makes a concession. 'Never mind about whether platonists can give an account of how we come to have mathematical knowledge,' he says. 'They can't even give an account of how it is that mathematicians' beliefs are usually true.' But, having met Field's challenge, the platonist is still not in the epistemological clear. It is as if someone defending Smith against a charge of murder were to say, 'Never mind about proving that Smith climbed the stairs and shot the victim. You can't even prove that he could have got out of his wheelchair!' If it is subsequently demonstrated that Smith could have got out of his wheelchair, his guilt is still far from being established. An account of how it is that someone's beliefs in a particular domain are true may fall short of an account of how it is that that person has knowledge of that domain, just as an account of how someone got out of his wheelchair may fall short of an account of how it is that that person climbed the stairs. I maintain that an account like Balaguer's does fall short.

Suppose, unbeknown to Daniel Defoe, that there really was a Robinson Crusoe who was shipwrecked on an island and who underwent adventures that exactly and completely match those undergone by the hero of Defoe's novel.[12] According to Balaguer, there is an important sense in which the novel truly describes part of the real world. Suppose someone, say Myrtle, were to read the novel and, being ignorant of its author's intentions, were to believe it to be true. In that case, it would be a fact that for any sentence p in the novel *Robinson Crusoe* if Myrtle believes that p, then p truly describes part of the real world. Furthermore, given that there really was a Robinson Crusoe who was shipwrecked, etc., we can account for that fact. Moreover, Myrtle's 'Robinson-Crusoe' beliefs are reliable, especially if she is a careful reader, has a good memory, and constantly refers back to the text. But do Myrtle's true beliefs constitute knowledge? Does she know that Robinson Crusoe and his island exist? No she does not. The concept of knowledge may be controversial, but if knowledge is more than mere true belief, then *prima facie* Myrtle lacks knowledge. The challenge for platonists is to convince us otherwise or to argue that Myrtle's case is not analogous to that of mathematicians and their supposedly true beliefs about the occupants of a plenitudinous, but platonic, realm.

A further, but related, objection can be put more briefly. Consider an even stronger conditional claim than Balaguer's:

(*)   If FBP is true, then mathematicians know that p (where 'p' is a mathematical statement).

Suppose we know (*) to be true. Should we then accept mathematical platonism? Of course not. According to mathematical platonism, there exists a realm of acausal objects and (*) gives no grounds for believing that such a realm exists, because (*) may be true, but its antecedent false.

12.6. PLATONISED NATURALISM AND PRINCIPLED PLATONISM

I turn to another platonism that is remarkable for its ontological profligacy. The platonism of Bernard Linsky and Edward Zalta (1995) also asserts a plenitude of abstract objects, but it differs from Mark Balaguer's full-

---

[12] The adventures of Alexander Selkirk only roughly match those of Robinson Crusoe whilst I am entertaining an exact match between novel and reality. Paul Griffiths suggests that a more realistic example would be Defoe's *Memoirs of a Cavalier*, which, at the time of its publication, was widely believed to be a true and accurate record of actual events. I choose (and stick with) *Robinson Crusoe* because the remoteness of its island is somewhat analogous with the remoteness of the platonic realm (and the novel is better known).

blooded platonism in a number of important respects. However, its supposed epistemological virtues are subject to similar objections.

Linsky and Zalta call their platonism a *principled* platonism (or PP) and place it within an overall position, which they characterise as *platonised naturalism* (as opposed to the naturalised platonisms of Quine, Maddy, Armstrong and others).

They begin by distinguishing their position from other platonisms. According to Linsky & Zalta, most platonisms model their abstract objects too closely on physical objects (1995, p. 532). Physical objects are sparse, complete and subject to an appearance/reality distinction. They are sparse because there are possible physical objects that do not exist as concrete objects, they are complete because those that do exist are determinate down to the last detail, and they are subject to an appearance/reality distinction because at least some of those details cannot be known in advance of empirical enquiry. Abstract objects conceived of on this model run into familiar epistemological difficulties. If abstract objects are 'out there in this sparse way', it is a puzzle how we can come to have knowledge of which ones exist and what properties they have.

Linsky & Zalta conceive of abstract objects in a fundamentally different way. They claim that, on their conception, platonism can be seen to be compatible with an orthodox naturalistic position. According to PP, there is a plenitude of abstract objects. Unlike FBP, this includes both complete and incomplete objects. A complete object is one that, for every property there is, has either that property or the negation of that property. Furthermore, abstract objects have only those properties that specify them and, accordingly, they have no hidden properties. But abstract objects do not 'have' their specified properties in the usual sense. Rather, they *encode* their properties.

The notion of *encoding* is the most distinctive aspect of PP. It is a primitive notion that provides an alternative reading of 'Object x has property F'. On the traditional reading, an object x exemplifies F, while on the alternative reading, an abstract object x may encode F. The notion of encoding is governed by three main principles:

(E1)   A comprehension principle that states that for every condition on properties, there is an abstract individual that *encodes* exactly the properties satisfying the condition.

(E2)   A necessity principle that states that if an object possibly encodes a property, then it does so necessarily.

(E3)   An identity principle that states that abstract objects are identical if and only if they encode the same properties.

The comprehension principle asserts the existence of a plenitude of abstract objects, some complete and others incomplete with respect to the properties they encode. For example, there is a complete abstract object that encodes all and only those properties that John Stuart Mill exemplifies. There is an incomplete abstract object that encodes only the properties of being dead, white and male. That object does not encode the property of being English, nor does it encode the property of not being English. Talk of completeness and incompleteness here is a little misleading. These abstract objects are only complete or incomplete with respect to the properties they encode, not with respect to the properties that they exemplify. All objects are complete with respect to the properties they exemplify. Thus the 'dead-white-male' abstract object exemplifies the property of not being English (and of not being German or Chinese and so on) and also the properties of not being dead, not being white and not being male. It does however exemplify the properties of being abstract, acausal, and mentioned in this book.

We begin to see how the notion of encoding together with the comprehension principle may give us an epistemic handle on abstract objects. If abstract objects are as PP proposes, then it seems that 'no matter what properties one brings to mind to conceive of a thing, there is something that encodes just the properties involved in that conception' (Linsky & Zalta, 1995, p. 537).

PP is intended to apply across the board to any discourse in which abstract entities rear their heads. PP-ists claim that their theory provides analyses of properties, propositions, possible worlds, fictions, etc. I shall focus on the PP-ist analysis of mathematical theories and objects.

This analysis first requires the notion of an object encoding a proposition. An object x encodes the proposition p by encoding the property of being such that p.[13] Then, according to PP, a mathematical theory T is that abstract object which encodes just the propositions asserted by T. This may seem circular, but it isn't. According to PP, for any given set of (purely mathematical) propositions there is a unique abstract object that encodes just those propositions. Traditionally, mathematicians and logicians have called the set

---

[13] Note that any object *exemplifies* the property of being such that p iff p. Hartley Slater complains that the suggestion that there are such properties is a grammatical nonsense. He demands to know what kind of an English sentence is 'Mary is such that George loves Barbara' (Slater 1997, p. 169). I shall let this pass. Linsky & Zalta do at least give truth conditions for the exemplification of such properties.

of propositions in question (or the conjunction of the propositions) a theory. Instead, PP-ists identify a theory with the abstract object that encodes the members of such a set. The members of the set are closed under the consequence relation, so that a theory encodes all of its theorems, not just its axioms.[14] Truth in a theory is equivalent to being encoded by that theory. In other words, proposition p is true in theory T iff T encodes the property of being such that p.

With this apparatus in place, the PP-ist can now give an account of the mathematical objects of a given theory T. The mathematical object ß of theory T is that abstract object that encodes just the properties F such that it is true in theory T that ß exemplifies F. For example, the number 1 of Peano number theory (PNT) is the abstract object that encodes exactly those properties that, according to PNT, the number 1 exemplifies. Thus, the claim that ß exemplifies F in T is equivalent to the claim that ß encodes F.

Linsky & Zalta claim that their principled platonism avoids Benacerraf's challenge (1995, p. 546ff). The idea is that knowledge of and reference to abstract objects is based on descriptions alone. It follows from their comprehension and identity principles that for any description there is a unique abstract object that encodes just the properties referred to in the description.

> All one has to do to become...acquainted de re with an abstract object is to understand its descriptive, defining condition, for the properties that an abstract object encodes are precisely those expressed by [its] defining condition.' (p. 547).

Their epistemology raises two problems. Can we always successfully refer to the properties in question, and how do we know that their governing principles, in particular, the comprehension principle, are true?

12.7. REFERRING TO THE OBJECTS OF PP

PP does not appear to face a uniqueness problem. According to (E3), one and only one object encodes the properties that are exemplified, for example, by the number 1 of PNT. So, if (E3) is true, there is no uniqueness problem. How we know that (E3) itself is true is discussed in the next section.

At first glance, it may appear that Linsky & Zalta's principled platonism faces similar problems to Balaguer's full-blooded platonism when it comes

---

[14] According to PP, there is an abstract object that encodes just the axioms of T, but the closure condition ensures that that object is not a theory.

to providing a convincing account of how we successfully refer to mathematical objects. One might argue as follows. Consider the axioms of djinn arithmetic (D1-5). These axioms (and their consequences) tell us all there is to know about djinns and their properties. Take, for example, the property of being the meta of Genie. No object exemplifies that property. But what would have to be the case in order that an object did exemplify that property? There seems to be no answer to that question, because there is no such property (just as there is no such property as the property of being the Jabberwock). The fact that the terms of the theory are effectively connotation-free ensures that this is so. And if there is no such property, then no object can encode that property.

Nor can there be any objects that encode the theorems of djinn arithmetic. Consider p_1, the proposition that the meta of Genie is a djinn. p_1 certainly appears to be derivable from (D1) and (D2), but what would it take for an object to exemplify the property of being such that p_1? Once again, there is no answer. And the same, it would seem, can be said of pure mathematical theories such as Peano Number Theory.

As Linsky & Zalta note, the epistemology for their abstract objects 'depends, of course, on the fact that we can refer to the properties involved in the description' (1995, fn. 48). If there really are abstract objects that encode ordinary properties, then it appears that we can refer to those objects because we can refer to ordinary properties. But the properties involved in describing djinn and Peano arithmetic are so 'thin' and indeterminate that it is doubtful that there are such properties let alone that we can refer to them.

As I say, one might so argue. But to do so would be to overlook further resources available to PP-ists. Zalta (1983 & 1988) has developed his theory of abstract objects to encompass abstract properties and relations. Briefly, abstract properties and abstract relations are treated as abstract entities that encode properties of properties and properties of relations (1983, p. 160). Just as abstract objects may encode properties that are 'incomplete', so abstract properties and relations may encode higher-order properties that are 'indeterminate'. Mathematical properties and relations are abstract. They encode just the properties of the properties and relations that are attributed by the mathematical theory in question. Clearly, djinn arithmetic is simply a notational variant of PNT. The meta and successor relations play the same role in each theory and, according to PP-ists, there is nothing more to those relations than the role ascribed to them by the theories. Hence, the terms 'meta' and 'successor' refer to one and the same abstract entity, the one that encodes the property of playing that role (Linsky & Zalta 1995, p.543-45). So PP-ists can agree that although nothing could exemplify the property of being the successor of Zero (or the meta of Genie), they may still insist that there is an entity that encodes that property.

Some may see this response as the piling of metaphysical weirdness on metaphysical weirdness but that is not, on the face of it, an epistemological objection so I do not pursue it here.

12.8. KNOWLEDGE OF THE PP PRINCIPLES

As with full-blooded platonism, even if we suppose that the PP-ists can provide a convincing account of how we might refer to the inhabitants of their abstract realm, we may still wonder how we could know that such objects exist. Linsky & Zalta recognise this concern. According to them, we can know of such objects' existence from our knowledge of the governing principles of PP, in particular, the comprehension principle, which states that a plenitude of such objects does exist. They claim that '[t]he comprehension principle as a whole…is synthetic and known *a priori*' (1995, p. 547). As an existence claim, the principle is clearly synthetic. They concede that if it is to be known, it must be known *a priori* because it is not subject to con-formation or refutation on the basis of empirical evidence. But how can such a claim be known *a priori*, especially within the constraints of a naturalised epistemology? They argue that such truths can be known *a priori* 'if they are required for our very understanding of…naturalistic theories' (1995 p. 548).

The comprehension principle is a central component of Zalta's inten-sional logic with its accompanying metaphysics (Zalta 1988). In other words, it is central to the logic of encoding as sketched above. According to Linsky & Zalta, this logic provides the best means for analysing and explain-ing problematic intensional constructions, mathematical language, and much else that is ubiquitous to scientific theories. A particular strength is that its ontological profligacy ensures that it is applicable to all possible scientific theories.

The ingenuity and comprehensiveness of Zalta's intensional logic cannot be denied, but the claim that it is the best on offer is bold and has not been met with widespread agreement.[15] But even if we agree that it is the best available explanation, it does not follow that its fundamental principles are (or even may be) known *a priori*. It is easy to imagine situations in which it would not be rational to believe our best available theory. This is one. Elsewhere, I have argued that we cannot have knowledge of the existence of

---

[15] Menzel (1993), Deutsch (1993), and Anderson (1993) make a number of criticisms to which Zalta (1993) replies. See also Slater (1997).

acausal objects. Furthermore, the comprehension principle appears to be a claim for the necessary existence of abstract objects, but the notion of necessary existence is itself a dubious one. Most striking is the thoroughly *ad hoc* air of this 'best' explanation. The existence of this abstract realm is offered as an explanation of our mathematical (and other) language and activities. But its existence has no empirical consequences. So it is difficult to see how the theory would lack any of its effectiveness if the abstract realm simply does not exist. Against these problems we have only the dubious epistemological principle that belief in the best available theory constitutes *a priori* knowledge of that theory.

Although a plenitudinous realm of abstract objects appears to offer epistemological advantages, the causal isolation of such a realm means that reference to and knowledge of its inhabitants is as problematic as ever. Neither Balaguer's full-blooded platonism nor Linsky & Zalta's principled platonism comes with a viable platonist epistemology.

CONCLUSION

In my introductory chapter, I noted John Burgess's complaint that there has been no sufficiently detailed defence of the causal objection to platonic knowledge (Section 1.5). In particular, he complains that Benacerraf's 1972 sketch of the objection remains 'the most detailed in the literature' (Burgess 1990, p. 3). Perhaps he overlooked a paper by Penelope Maddy (1984) that examines the issue with some care, although, to be fair, her defence of the objection is somewhat equivocal. More recent work by Hartry Field (1989, pp. 25-30) and Albert Casullo (1992) also presents versions of the anti-platonist case with vigour and in some detail.[1] Even more recently, Burgess himself, with Gideon Rosen, presents a balanced and detailed assessment of the epistemological difficulties for mathematical platonism (Burgess & Rosen 1997, pp. 26-60). My book-length defence of the causal objection finally lays Burgess's original charge to rest. The causal objection to platonic knowledge has now been argued in detail.

However, this book does more than show that the causality objection can be argued in detail. It also shows that the objection presents serious difficulties for the platonist position and cannot easily be dismissed. The many and various counterexamples to a causal constraint on knowledge (in particular, on existential knowledge) can be dealt with, either by showing that they are unsuccessful, or by modifying the constraint to avoid them. Furthermore, a positive case can be made for the necessity of a causal constraint in any plausible theory of knowledge. Finally, platonists cannot provide a convincing account of how platonic knowledge could be obtained by human knowers. In the absence of such an account, we should remain sceptical of the claim that platonic entities exist.

---

[1] See also Maddy (1990) pp. 36-50.

Crispin Wright (1988) has claimed that the 'defeat [of the causality objection] is firmly in prospect' (p. 438, fn. 30). Wright is wrong. Such a defeat has not even appeared on the distant horizon. It is epistemological platonism, the claim that we can have knowledge of platonic entities that is looking shaky. And if epistemological platonism falls, ontological platonism, the claim that platonic entities exist, must fall with it.

APPENDIX I

DOES KNOWLEDGE REQUIRE BELIEF?

I.1. THE TRADITIONAL ANALYSIS OF KNOWLEDGE

The strong causal condition on knowledge states that:

(SC)   S knows that p only if the fact that p is causally connected to S's belief that p.

(SC) entails that knowledge requires both truth and belief. This accords with what has become known as the traditional analysis of knowledge:

(TK)   S knows (at t) that p if and only if:
        (a) p is true
        (b) S believes (at t) that p
        (c) S is justified (at t) in believing that p. (Gettier 1963, p. 121)

In this appendix, I briefly consider the truth condition and then examine the belief condition in detail.

I.2. THE TRUTH CONDITION

Occasionally, a layperson may argue that a false proposition could be known. His example is typically that in the past people knew that the earth was flat, although we now know that it is not flat. But further questioning usually reveals his position to be truth-relativism. He argues that the ancients

196

knew that the earth was flat because it was true for them. So it turns out that he does take truth to be a condition for knowledge, albeit relative truth.

There are examples of ordinary language uses of 'know' and 'knowledge' where truth is absent or appears to be absent. We may speak of the scientific knowledge of the past although much, even most, of its content is false. This is best seen as an ellipsis for what scientists in the past took to be knowledge or, perhaps, what they were most justified in believing to be knowledge. Proponents of the 'sociology of knowledge' are particularly prone to use 'knowledge' in this elliptical sense (Brown 1989, p. 180). Similarly, we may speak of someone's being knowledgeable in the mystic arts or in alchemy. This manner of speaking is elliptic or idiomatic. We could mean that such persons understand and adhere to the corpus of beliefs that is taken to be knowledge by such a school of thought. In another context we could mean that they know (*really* know) what the corpus of beliefs is that is held by adherents to such a school of thought. In other words, they have knowledge *about* the occult arts. Or we may mean that they are skilled in the practices of the art. They have knowledge-how rather than knowledge-that.

There is also an ironic use of 'know' (Ginet 1975, pp. 12-13). We might say that the Australians *knew* they would win when they took the field against the New Zealand cricket team although, in fact, they were to be defeated. But here our meaning is not literal. We simply wish to emphasize our belief that the Australians were overconfident.

Such examples need not detain us further. We require, not a list of various shades of meaning, but a robust notion of knowledge. Such a notion must be useful, internally consistent, and draw as much as possible on what is known about human knowers and the world they inhabit. On such a notion it is impossible for someone to know something that is not the case.

I.3. THE CONSISTENCY OF BELIEF AND KNOWLEDGE

The belief condition has been challenged in two ways. The first way is to point to certain linguistic facts, which purportedly indicate that knowing that *p* is inconsistent with believing that *p*. The other way is to produce counter-examples in which it is claimed that someone knows something without believing it.

Are knowledge and belief inconsistent? To say 'I do not believe that; I know it' is apparently acceptable and meaningful. But it doesn't follow that one can know without believing, let alone that one cannot know unless one does not believe. After all, it is also acceptable and meaningful to say 'I do not just believe that; I also know it'. Indeed, reflection indicates that the former statement is simply a more emphatic way of stating the latter. It is analogous to our saying 'That is not a house; it is a mansion', when we mean, emphatically, 'That is not just a house; it is a mansion' (Lehrer 1974, p. 50). This sort of confusion arises when we don't distinguish between what a speaker implies and what is implied by what the speaker says (see Grice 1957). Often they are the same thing, but not always. In the example, the speaker's statement (what the speaker says) implies that he does not believe, but what he is implying is that his state consists of more than mere belief. Once we make this distinction, we have no reason for supposing that knowing and believing are mutually exclusive.

I.4. RADFORD'S COUNTEREXAMPLE

A good example of an attempt to provide a counterexample to the belief condition is Radford's (1966) history test story. Jean (a French Canadian) protests that he knows no English history but his friend Tom insists on quizzing him nonetheless. Jean just guesses and gets the answers hopelessly wrong until, to his surprise, he is right about the dates of the deaths of Elizabeth I and James I. The explanation for Jean's success is that in the past he was made to learn those dates but has subsequently forgotten the episode. Although he is sure that he is guessing, he is in fact remembering correctly. Should we then agree with Radford that Jean knows that Elizabeth died in 1603 although he does not believe that she died in 1603?

There are two ways in which Radford's claim could be rejected. We could argue either that Jean does not know or that he in fact believes. What reasons can we give for saying that Jean does not know, apart from the fact that he does not believe (which would beg the question in favour of the belief condition)? First we should note that in quiz contexts, it is quite usual to say that someone knows the correct answer simply on the grounds that she has given the correct answer. Suppose the quizmaster announces that Sarah will win the car if she knows when Elizabeth I died. Sarah has no idea (and never has had any idea) but guesses that it was 1603. Sarah knows the correct answer and so she wins the car even if she admits that it was only a lucky guess. But if we ask Sarah if she really knew that Elizabeth died in 1603, she is most likely to say 'No, I only guessed that'. The fact that Sarah or Jean knows the correct answer (in a quiz context) does not imply that he or she knows that Elizabeth died in 1603.

But there is a difference in the case of Jean. Although he also believes that his answer was correct by chance, it was actually correct because he remembered what he had once learned. In cases of someone learning that *p* and later remembering that *p* we usually ascribe knowledge. This feature is present in Jean's case and this explains our inclination to see it also as a case of knowledge. But Jean's case is a borderline case of knowledge, in the sense that it is on or near the border between knowledge and non-knowledge. It is borderline because it lacks other features that we usually associate with knowledge. In straightforward cases, the knower is convinced of the truth of *p* and is prepared to say so. Jean lacks this conviction. Now borderline cases are common for most terms in everyday use. It seems that there is a sense in which Jean knows that Elizabeth died in 1603 and a sense in which he doesn't know. If it is a genuine borderline case then no amount of argument or further evidence will decide whether or not the term should apply.

Suppose that it is a genuine borderline case. Must we then agree that there can be borderline cases of knowledge that do not meet the belief condition? Before so agreeing, we should examine the assumption that Jean does not believe that Elizabeth died in 1603. In cases where someone says that *p* because he remembers that *p* we would usually ascribe belief. This feature is present in Jean's case. However, a feature which is lacking, but which is present in straightforward cases of belief, is that Jean is not convinced of the truth of his answer and is not prepared to assert that it is true. In other words, Jean lacks conviction. So, it appears that Jean's belief is a borderline case of believing, and that it is borderline for the same reason that his knowing was borderline.

Another possibility is that Jean's belief or near-belief is borderline because he is not aware of having a belief. I have already argued that conscious awareness is not a necessary condition for belief (Section 2.1), but we do expect someone to be aware of her beliefs about some issue when that issue is raised. On the face of it, if we are prepared to accept that beliefs can be unconscious, there is no reason why we should deny that possibility when certain circumstances obtain, such as the issue being brought to the subject's attention. Furthermore, lack of awareness is clearly not a bar to Radford's ascribing knowledge to Jean.

There is an asymmetry in Radford's ascription of knowledge and of non-belief. He suggests that Jean lacks belief because he lacks conviction or awareness but that he knows (albeit marginally) in spite of that same lack of conviction or awareness. We can agree that there may be no 'fact of the matter' in borderline cases, but still insist on consistency of application. If Jean is denied both belief and knowledge because he lacks conviction or awareness, or if he is ascribed both belief and knowledge in spite of that lack of conviction or awareness, then in neither case is the belief condition violated. What is required from Radford is an argument for treating the two ascriptions differently, and this is what is lacking.[1]

Is there an argument for treating knowledge and belief differently with respect to conviction or awareness? One intuition is that belief requires less conviction than knowledge. Suppose Jane knows that p and is strongly convinced that p and that Jim is in exactly the same epistemic position as Jane except that he is rather diffident in his assertion that p. Although we are inclined to say that Jim also believes that p (albeit less strongly than Jane), we are not so inclined to say that he also knows that p.[2] If this is correct, then it counts strongly against the history-test example. If Jean doesn't have enough conviction for belief, then he has insufficient conviction for knowledge.

Can we point to intuitions that pull in the opposite direction? Examples of unwitting seers are thought to do so. D.S. Mannison (1976) provides an example. Suppose that whenever Harry goes to a race meeting, his friends ask him to pick a winner for each race. He protests (truly) that he knows nothing about horses and racing, and names horses only to be sociable. His horses always win (pp. 141-42).[3] A mysterious underlying mechanism accounts for this phenomenon.[4] Mannison's claim is that when Harry makes his choice he doesn't believe the horse will win but he does know which horse will win.[5]

We can make the same response to this example as to the history test example. If there is a minimal sense in which Harry knows, then there is also a minimal sense in which he believes. But we can attempt to undermine this response by altering the example so that Harry is merely asked (say as part of an experiment for which he is paid) if he has anything to say about the forthcoming race. Each time Harry says something like 'I wonder whether

---

[1] This argument draws on Armstrong (1969-70) pp. 29-33 and Lehrer (1974) pp. 61-63.

[2] Plantinga (1993) favours this intuition to the extent that under his conditions for knowledge 'the more firmly S believes B the more warrant B has for S' (p. 19).

[3] Compare with D.H. Lawrence's 'The Rocking Horse Winner'.

[4] We need not suppose that the mechanism involves anything as unlikely as backwards causation. Either whatever causes a certain horse to be the winner also causes Harry to name that horse, or else Harry's naming the horse causes it to win. Less bizarre examples are possible incorporating the mechanisms of subliminal messages or hypnotism.

[5] What sort of knowledge Harry has is debatable. If he knows *which* horse will win is this equivalent to, or does it entail, his knowing *that* horse n will win? Or does he only know *how* to pick winners? I shall argue that he lacks propositional knowledge. If that involves denying that he has knowledge-which, so be it.

horse n might win' and the horse about which he muses always wins (Shope 1983, p. 186). The suggestion here is that there is not even a minimal sense in which Harry believes because he never picks any horse as a winner, but the situation with respect to his 'knowing' which horse will win has not changed. A further alteration to the example along similar lines shows that the intuition to ascribe knowledge must be mistaken. Suppose Harry's response each time is to say 'I don't think horse n will win', but the horse that he mentions is always the winner. To suggest that Harry still knows which horse will win is surely perverse, even though the same underlying mechanism is operating as reliably as ever. And if he does not know in this case, then he does not know in the previous two scenarios.

I.5. INFORMATION AND KNOWLEDGE

How does the mistaken intuition arise? It arises because of confusion between information and knowledge. A calculator can tell us the square root of 529. But it does not know that the square root of 529 is 23, any more than a book of mathematical tables knows. The calculator and the book can possess and provide the information that the square root of 529 is 23 and by extracting that information we can come to know what the square root of 529 is. So we can speak of the calculator as providing us with knowledge and this gives rise to talk of the calculator's possessing knowledge. Similarly we speak of books or computer databases as storehouses of knowledge. But this is a derivative sense of knowledge and it in no way sanctions a slide to asserting that books and databases know their contents. Harry possesses the information as to which horse will win. Thus he can provide us with the knowledge as to which horse will win. But it is not the case that he, himself, knows which horse will win.

But that still does not explain the inclination to say that he knows in the first, and perhaps second case, but not the third. That inclination arises because he behaves in the first case as if he believes that the horse will win (he picks it) and in the second case as if he hypothesises that it will win and a hypothesis may be seen as a sort of proto-belief. If someone behaves as if he were in state k then we are likely to ascribe state k to him. But if he is not really in state k then our ascription is mistaken although his behaviour may be sufficient for the ascription to be borderline or analogous. Harry behaves as if he believes that a certain horse will win (although we know he doesn't) and he has the capacity to provide us with knowledge of which horse will win. This is sufficient to account for the mistaken inclination to say that he knows. Remove the behaviour (and have him behave as if he doesn't believe, as in the third case in which he predicts that the horse will lose) and it is clear that he lacks knowledge except in a derivative sense. It is also clear that without an attitude of belief toward a proposition, we do not have a case of genuine knowledge of that proposition.

There are no grounds for denying the truth condition or the belief condition on knowledge. The strong causal condition on knowledge is not undermined by challenges to those conditions.

APPENDIX II

CAN WE CHOOSE OUR BELIEFS?

II.1. BELIEF-VOLUNTARISM AND EPISTEMIC JUSTIFICATION

Belief-voluntarism is the doctrine that we can choose at least some of our beliefs. Many philosophers and theologians in the past have been belief-voluntarists, for example, Augustine, Aquinas, Descartes, Kierkegaard and Roderick Chisholm. The doctrine is important (and probably necessary) for their deontological conception of epistemic justification. They claim that a subject in certain epistemic situations ought, or ought not, to have certain beliefs. Because of the principle 'Ought implies can', such a claim could only be upheld if belief-voluntarism were true. But the issue of the correctness of the deontological conception of epistemic justification need not concern us here. What is at issue is the doctrine of belief-voluntarism itself.

II.2. BELIEF-VOLUNTARISM — IS IT POSSIBLE?

An early opponent of belief-voluntarism was David Hume:

> We may, therefore, conclude, that belief consists merely in a certain feeling or sentiment; in something, that depends not on the will, but must arise from certain determinate causes and principles, of which we are not masters. (1739-40/1978, p. 624)

Presumably, Hume thought that it was a contingent truth that humans do not form beliefs at will. Although he claims that we are not masters of those 'determinate causes and principles', he does not claim that it is impossible that we be their masters. Bernard Williams (1972, pp. 148-51) thinks that it is not merely a contingent truth that we cannot believe something at will, but a necessary one. He argues that belief-voluntarism is logically impossible. He has two arguments. The first stems from his claim that a characteristic of beliefs is that they aim at truth; to believe something is to take it as being true. He also claims that no one could have the ability to acquire beliefs at will without realising that he has that ability. How (asks Williams) could I take something to be true, while believing that I acquired it at will? Such a 'belief' would lack the truth-aiming characteristic, thus it would not be a genuine belief. So for any belief that I acquire at will it must be the case that I could not believe (at the same time, in full consciousness) that I acquired it at will. But then I could not believe that I could acquire beliefs at will. But believing that I can acquire beliefs at will is a necessary condition for being able to acquire beliefs at will. Therefore, it is impossible for anyone to acquire beliefs at will.

I shall examine two of the claims on which Williams's argument depends:

(1)   Necessarily, one cannot believe that p and (at the same time, in full consciousness) believe that the belief that p was acquired at will.

(2)   Necessarily, if one can acquire beliefs at will, then one believes that one can acquire beliefs at will.

Both of these claims can be challenged. (1) is supposed to follow from the truth-aiming characteristic of belief. If I believe that p then I must believe that p is true. So if I acquire a 'belief' by an act of my will, I couldn't possibly regard it as something I take to be true, because I know that truth was not a consideration in its acquisition. But we can agree that beliefs necessarily aim at truth without drawing this strong conclusion. Belief-voluntarism is not the thesis that we can choose to believe anything at all. Beliefs chosen with complete disregard for the evidence, or in the absence of any evidence, would no doubt conflict with the truth-aiming characteristic. But what of cases where there is good evidence for p and for q, where p and q are incompatible, and where the issue is important to the potential believer. These are the sort of cases that interest belief-voluntarists. And such cases are prevalent in philosophy. For example, the existence or non-existence of God, or mind dualism $v.$ materialism, or even freewill $v.$ determinism. They also arise in many other circumstances, such as when a scientist must choose between competing theories, or when someone lost in the bush is faced with alternative tracks. In such circumstances, believing that one has freely chosen to believe that p is perfectly compatible with believing that p is true. If the truth-aiming characteristic of belief were so strong as to make those two beliefs incompatible, then it would seem be strong enough to make the belief that one is fallible (with respect to beliefs) incompatible with any belief. But there is no problem with believing that p is true at the same time as believing that it could be false. Similarly, there is no problem with believing that p is true at the same time as believing that the belief that p was acquired at will (cf. Govier 1976, p. 647). So (1) does not follow from the truth-aiming characteristic.

Williams might respond that his argument is aimed against the possibility of voluntary belief where there is a complete disregard for the truth, and that in the cases mentioned above there is clearly a regard for the truth and therefore they are not examples of genuine ('full-blooded') voluntary belief. But to argue so would be to considerably restrict his position.

Moreover, there is another, more direct, objection to (1). We can devise counterexamples. Although I may believe that I acquired at will the belief that p with total disregard for the truth, I may now believe that I believe that p for good reason. For example, I may now believe that last night I decided (without regard to the truth) to believe that my neighbour's new letterbox is blue. I may also now believe that that my neighbour's new letterbox is blue and I believe this because I can see that it is blue. There is nothing contradictory about those two beliefs. A rational person could hold them both in full consciousness at the same time. So (1) is false.

Barbara Winters (1979, p. 253) claims that if we amend (1) by replacing 'was acquired' by 'is sustained' then Williams's original argument is no longer valid for it will now no longer yield a conclusion about belief acquisition. But if the replacement were 'both was acquired and is sustained', I suspect that a valid argument to the conclusion that it is impossible for anyone to both acquire and sustain beliefs at will could be devised. However, this would be a further restriction of Williams's position.

Williams's claim (2) that necessarily, if one can acquire beliefs at will, then one believes that one can acquire beliefs at will, is also implausible. Either it is supposed to follow from a more general principle for voluntary action or from some property peculiar to belief acquisition. Presumably, the general principle is that if one can perform an action at will, then one must believe that one can perform such an action at will. This is clearly false. Someone could have the ability to wriggle his ears at will without ever suspecting that he has that ability. If you don't try, you never know what you can do! Even if one has exercised such an ability, it does not follow that one must now believe that one has the ability. We can be notoriously mistaken about the causes of our past actions. There is the possibility of mis-remembering. Or consider someone who is pathologically lacking in confidence. No sooner does she successfully perform an action at will than she begins to doubt that it was achieved through her own efforts. Williams does not give any reason for thinking that voluntary belief acquisition must be a special case and it is not clear what reasons could be adduced for regarding it as such (cf. Winters 1979, p. 255).

Williams does have a second argument for his claim that acquiring beliefs at will is logically impossible. He claims that central to our concept of empirical belief is the idea of a connection between the environment, the believer's perceptions and the belief that results. In other words, acquisition of an empirical belief involves the notion of coming to believe that something is so because it is so. Since 'beliefs' acquired at will could not satisfy the demands of this notion, they could not be empirical beliefs. This sounds alarmingly like a 'strong causal theory of empirical belief'. Is Williams claiming that S empirically believes that p only if the fact that p causes S's belief that p? Surely not, for that would rule out the possibility of false empirical beliefs. Perhaps he is merely claiming that S could not believe that his belief is an empirical belief unless he believed there was a causal connection between fact and belief via his perceptions and that he could not believe that of a belief acquired at will. This claim raises the same problems as the one discussed immediately above. It is logically possible to have false beliefs about our beliefs. Besides, whatever Williams's concept of empirical belief may be, it does not rule out the possibility of 'non-empirical' beliefs about the world which are acquired by divine intervention, by neuro-surgical intervention, innately, from dreams, by free choice or even *ex nihilo*. Such 'beliefs' could meet criteria central to the concept of belief, that of being a propositional attitude which gives rise to a disposition for certain behaviour and/or provides a rationalisation for action. And if such beliefs were introspectively indistinguishable from beliefs acquired percept-ually, then there seems no reason to deny that they would fall under our usual concept of belief. Since we often completely forget the origins of our beliefs, such introspective indistinguishability is clearly possible. Williams is conflating the notion of a belief acquired through sense experience with the notion of a belief about the world. Although the expression 'empirical belief' is ambiguous between those two notions, trading on that ambiguity is not enough to convince us that it is logically impossible for there to be beliefs about the world that are acquired at will (cf. Govier 1976, pp. 647-48).

Williams's arguments do not establish his thesis that voluntary belief is logically impossible. Although it does not follow from this that his thesis is false, it does follow that we have no good reason to accept it. Jonathan Bennett (1990) claims to have a strong intuition that belief-voluntarism cannot possibly be true. He attempts to establish this. But in the end he admits defeat.

I turn, therefore, to the Humean view that it is a contingent fact that human beings do not acquire beliefs at will. To begin with I shall be concerned with whether or not we have direct or immediate or basic control over our beliefs by the exercise of choice, in the same way that we have direct or immediate or basic control over our bodily movements by the exercise of choice. Perhaps our voluntary control of our bodily movements is indirect. Perhaps when I choose to raise my arm a number of bodily (and perhaps mental) processes must mediate between my choice and the raising of my arm. By 'direct' control I really mean control that is at least as direct as that which is involved in the voluntary control of our bodily movements.

II.3. DIRECT BELIEF-VOLUNTARISM — IS IT ACTUAL?

Consider how we usually describe perceptual belief acquisition. I turn around and see a book on the shelf. I acquire the belief that there is a book on the shelf. Would we not say that seeing the book caused me to believe that there is a book on the shelf, rather than that on seeing the book I chose to believe that there is a book on the shelf? The latter description sounds absurd, and it sounds absurd because it is absurd. It simply does not describe the situation as it actually occurs. The same applies in the case of (so-called) non-perceptual beliefs, for example, beliefs acquired from books, teachers, parents, etc. I do not read a passage from a book that I regard as reliable and then decide to believe what it says. It is my reading of the book that causes me to acquire the belief. Our beliefs are formed as a result of what seems to us to be the case or by how things strike us. We have no choice over how things seem to be or how they strike us.

Of course, I need not acquire the belief that there is a book on seeing the book. Other possible beliefs are congenial with my seeing the book. By saying that a possible belief is 'congenial' with my perceptual experience, I mean (roughly) that if the content of the belief were true then my perceptual experience would be indistinguishable from my actual experience. For example, my belief that Tim is in front of me may be congenial with the experience of my seeing his identical twin brother Tom in front of me. (Compare with Kripke's notion of an epistemic counterpart (1980, pp. 103-04), or Goldman's notion of a relevant alternative (1976, pp. 779-80). It might be argued that, although I cannot choose to believe just *anything* when undergoing that experience, I am free to choose from those beliefs that are congenial with the experience. The belief that there is a raven on the shelf would not be congenial with the perceptual experience I have when looking at a book (under normal conditions), but the belief that there is a cigar box cunningly disguised as a book would be congenial. Why could I not choose to believe the latter? Here I offer a Humean challenge. Try looking at a book and forming the belief that it is a cigar box. It cannot be done—not by one's own choice at any rate. Of course, if you have a friend who is given to leaving 'book-like' cigar boxes about the place, then you might suspect, even believe, that you are looking at such a counterfeit. But once again it will not be a matter of choice. You may discover on looking at the object on the shelf that you believe that it is a cigar box. But you will not have chosen to believe that it is a cigar box. That belief will have been caused (in part) by your knowledge of your friend's earlier eccentric behaviour that will in turn have been caused by that behaviour.

What about a situation where there is conflicting evidence as to whether it is a book or a cigar box? Suppose there is good but inconclusive evidence for each possibility and that the evidence is evenly balanced. Suppose, further, that it is important that you make up your mind as to which it is. Could you not, in that case, choose which to believe? Again the answer is 'No'. If neither alternative seems to you to be true, you cannot by an effort of will adopt either belief. What you really believe is that the evidence is evenly balanced. Similarly, when someone lost in the bush comes to a fork in the track, she would normally choose the path that she believes will lead to safety. But if each path seems to her to be equally likely, she will not choose to believe that one of them is the correct one. If it is better that she follow a path rather than stay where she is, then she will choose one path to follow and act as though she believed it to be the path to safety. The difference between believing and acting as though one believed is discussed further in the next section (Section II.4).

I offer another Humean challenge. Try choosing to believe, before it is tossed, that a fair coin will come down heads. No matter how much you try to convince yourself that you believe it will come down heads, in your 'heart of hearts' you will know that you still believe that it is just as likely to come down tails. This is not to deny that someone could not firmly believe that a tossed coin will come down heads. But it would not be a matter of choice. He might have a strong intuition or premonition that it will be heads. But we cannot choose to have such feelings. He might be influenced by the fact that the coin has come down tails five times in a row. But then the belief will have been caused by the run of tails plus a belief in the gambler's fallacy.

As mentioned above, belief-voluntarists usually do not claim that all beliefs are subject to the will. They are more likely to claim that it is only with respect to hypotheses that do not seem clearly true or false, particularly those concerning abstract matters of religion, philosophy, and pure science, that we have the ability to choose what to believe. For William James, it is only when the believer is confronted with a 'living, forced, and momentous option' that the will to believe is applicable (1897/1979, p. 14). By a living option he means one in which both of the hypotheses appeal as genuine possibilities, by a forced option he means one in which there are no alternatives apart from the hypotheses, and by a momentous option he means one for which the choice will have an irreversible and significant effect on the chooser's life. But my earlier comments apply equally to such options. It is not a question of how abstract or how far removed from immediate sensory perception that the hypotheses are. The choice between tracks in the bush, or even between book and cigar box may be just as living, forced, and momentous as whether or not to believe in God. In all such cases, the situation is basically the same. When we make a decision between hypotheses, it is either on the basis of a belief that has already formed or we choose without belief and proceed to act as if we had the appropriate belief.

I note in passing that it is questionable whether James was a genuine belief-voluntarist. In the face of criticism, he conceded that his famous essay 'The Will to Believe' might have been more felicitously entitled 'The *Right* to Believe' (1897/1979, pp. xxii-xxiii). For him, the main thrust of his thesis was that we have a right to believe in the face of inadequate evidence if and only if our option is living, forced, and momentous.

It may happen that after we have decided to act as if we believed something, psychological factors come into play causing the relevant belief to form. Human beings are egotistical creatures. Having adopted a position we tend, particularly if we have to defend it, to 'take it to heart' and thus come genuinely to believe it. On the other hand, we have all (I am sure) had the experience of adopting a position arbitrarily (when faced with incompatible hypotheses but inconclusive evidence) only to become quickly disenchanted. Once again, psychological factors have come into play, this time to cause disbelief. There seems no systematic way of predicting whether belief or disbelief will follow the adoption of a working hypothesis. Hence, it would be an unreliable technique for acquiring favoured beliefs. Even if we could perfect such a technique, it would involve choosing beliefs indirectly. I discuss indirect belief choice in Section II.5 below.

In the above examples, I have been concerned with the possibility of belief formation as the result of making a free decision. But consider the following kind of belief formation. In a recent edition of *Campus Review*, a survey of the research capabilities of Australian universities was published along with the reactions of some university vice-chancellors. Curiously, the vice-chancellors of those universities that came out top were reported as believing that the survey gave an accurate account of the situation while those vice-chancellors of universities that were placed towards the bottom of the list seemed to believe that the survey gave an inaccurate picture. Now, although we would not say that the vice-chancellors had *decided* to believe as they do, we might suggest that that they do believe what they want to believe, that they are indulging in 'wishful thinking'. Because wishful thinking is frowned upon, it must be something that the wishful thinker is responsible for and, therefore, there are grounds for supposing that it is an example of freely choosing one's beliefs.

The claim that wishful thinking is an example of freely choosing to believe assumes a compatibilist view of freewill. Compatibilists claim that an action is free if it is caused by the goals, desires and emotions of the agent. The vice-chancellors' beliefs were caused by their goals, desires and emotions, so wishful thinking apparently accords with this compatibilist notion of freewill. But it does not accord with the libertarian notion, since the beliefs in question have been fully determined by prior events and conditions. I have been arguing that we cannot freely believe and therefore, *a fortiori*, we do not have libertarian freewill with respect to our beliefs. The anti-causalist argument I discuss in Chapter 3 relies on libertarian freewill, so the weaker claim that we do not have libertarian freewill with respect to our beliefs will suffice since wishful thinking is not an example of libertarian freewill.

Although it is not necessary for my case, I am reluctant to relinquish my stronger claim that we cannot freely believe in either a compatibilist or a libertarian sense. In other words, I claim wishful thinking is not a case of voluntary belief even in the compatibilist sense. Space precludes a detailed defence of this claim, but I suggest that in the case of the vice-chancellors their desire to believe the best of their universities conflicts with their higher-order desires to want to believe only what is true. Sophisticated compatibilist theories (for example, Frankfurt 1971) require a match between lower and higher-order desires if it is to be a genuine case of freewill. On the other hand, if (heaven forbid) a vice-chancellor should have a higher-order desire to want to acquire beliefs irrespective of the truth, then this would put his belief formation far enough beyond the pale of rationality to disqualify his beliefs as being under his voluntary control.

One further point. Even if wishful thinking does involve libertarian free-will, beliefs that arise from wishful thinking would not qualify as items of knowledge in any case. So we have a further reason why examples of wishful thinking cannot advance the anti-causalist argument.

II.4. BELIEF AND ACCEPTANCE

I have argued that any apparent case of choosing to believe when one is faced with alternative, plausible hypotheses will be either a case of believing what seems be true (something over which one does not have control) or simply a case of behaving as if one believes. Paul Horwich (1991) argues that 'there is no genuine difference between *believing* a theory and being disposed to *use* it to make predictions, design experiments, and so on' (p. 1). Perhaps Horwich's argument applies to the distinction that I make between believing and behaving as if one believes. If I am making a distinction without a difference, then this may undermine my thesis of belief invol-untarism. I do not believe that I am making a distinction without a differ-ence, but even if I am, this does not undermine my thesis.

Horwich's target is scientific instrumentalism. He wishes to undermine the instrumentalist notion of 'acceptance' as van Fraassen (1980) employs it. It is this attitude of acceptance in an instrumentalist sense that he claims cannot be distinguished from the attitude of belief.

Horwich argues as follows:

> If we tried to formulate a psychological theory of the nature of belief, it would be plausible to treat beliefs as states with a particular causal role. This would consist in such features as generating certain predictions, prompting certain utterances, being caused by certain observations, entering in characteristic ways into inferential relations, playing a certain part in deliberation, and so on. But that is to define belief in exactly the way instrumentalists characterise acceptance. Thus, we have a *prima facie* case for the thesis that belief simply *is* acceptance. (1991, p. 3)

Horwich does not deny that we can treat theories instrumentally, if by this we mean employing them 'merely as heuristic devices for certain practical purposes.' (p. 4) He calls this 'local' instrumentalism. But he claims that acceptance (in the 'global' instrumentalist sense) involves much more than this. So much more that it is indistinguishable from a state of belief. Even if he is correct, this does not threaten my claim. If someone has acquired a disposition to use a theory and this amounts to his having acquired a belief in the theory, then this has no implications either way as to whether such an attitude to the theory has been freely chosen. We can choose to behave in a certain way. We can choose to behave in certain ways given certain circumstances. It is not at all clear that we can choose directly to acquire a certain disposition. There is no reason to give up my claim that we cannot choose to acquire a disposition that is indistinguishable from belief.

Besides, Horwich's argument is hardly convincing. According to the 'causal-role' theory of mental states that he affirms, mental states are to be individuated by their causes, effects, and interrelations with other mental states. Acceptance can be distinguished from belief on all three counts. Both may be caused by observations of the same type, but belief in a theory may (typically) be motivated by a desire to discover the underlying causes of those observations while acceptance may not. On the other hand, acceptance may be motivated by reading van Fraassen while belief may not.[1]

We can expect beliefs and acceptances to differ in relation to other mental states. Horwich suggests that instrumentalists will claim that true believers believe they believe whereas mere accepters do not (p. 3). But instrumentalists will claim more than this. They will claim that accepters believe that they do not believe while believers do not so believe. Horwich's response is that the difference arises, not from a difference in mental state, but from a confusion about that mental state on the part of the so-called 'accepter'. It is, he says, 'perfectly possible for someone to sincerely, yet mistakenly, deny her beliefs, that is to believe without believing she believes. So a difference between belief and acceptance is not implied merely by the possibility of conforming one's metabeliefs to instrumentalist doctrine.' (p. 4)

This talk of metabeliefs points to a *prima facie* case for supposing that belief is distinct from acceptance, a supposition that Horwich agrees is 'usually taken for granted' (p. 3). The distinction is based on introspection. It seems to us that we can clearly distinguish between those occasions when we genuinely believe and those when we merely act as if we believed. For those who believe that the deliverances of introspection are incorrigible this argument will be decisive. However, Horwich does not believe in introspective transparency and neither do I. I agree that we can be wrong about our mental states and that it is possible for someone to believe that she merely accepts a theory when in fact she believes it. But Horwich must

---

[1] Of course, both beliefs and acceptances could have all sorts of weird and wonderful causes. Here I mean rational beliefs and rational acceptances, as must Horwich. Or perhaps we are both concerned with typical causes rather than weird and wonderful ones.

claim that on every occasion that someone believes that she merely accepts a theory rather than believes it she is mistaken. Are our powers of introspection that fallible? Identity of belief and acceptance is not implied merely by the possibility that one's metabeliefs may be mistaken.

A belief and an acceptance may produce different effects. A belief that p will typically give rise to the utterance that p or 'I believe that p' while a mere acceptance that p will not. An acceptance that p will typically give rise to the utterance that 'I do not believe that p' or that 'We do not have good reason to believe that p' while a belief that p will not.

Horwich has not established that a genuine distinction cannot be made between belief and instrumentalist acceptance or any similar attitude. But he does indicate that the two attitudes are more similar than we might have suspected. This suggests another approach. Could it not be that anyone who is in a position to accept a theory (in the 'full-blooded' instrumentalist sense) would be psychologically unable to avoid believing the theory? Suppose someone were acquainted with the evidence for electron theory and had that theory explained to him to the extent that he believed that it is supported by all the evidence and that it is a useful predictor. Would not electron theory seem convincing to such a person? Perhaps not to every such person on every occasion. Only an extremely alert and strong-minded instrumentalist would be likely to resist for long.[2] So although acceptance and belief are distinct mental states, it is a contingent fact that the former is likely to be overtaken by the latter, and this contingent fact makes instrumentalism practically if not theoretically untenable. It also endorses the fact that we cannot choose what to believe.

II.5. CHOOSING BELIEFS INDIRECTLY

There is another way of forming beliefs that might be considered to be choosing to believe. Suppose it is observed that people exposed to a particular environment tend to form a particular belief. If I decided that I should like to have that belief, could I not then acquire it by choosing to expose myself to that environment? This is what Pascal had in mind when he suggested that by adopting a religious life one could come to believe sincerely in God. Having argued that one should believe in God for pragmatic reasons, he was worried that (as I have argued) one cannot actually choose to believe in God (1670/1966, pp. 150-52). He takes seriously the

---

[2] cf. Hume on his sceptical reflections (1739-40/1978, pp. 268-69).

objection that one may believe that one would be much better off believing that God exists and yet be unable to believe that God exists. He suggests that '...taking holy water, having masses said, and so on...will make you believe quite naturally' (p. 152). Note that Pascal (like James) had doubts about belief-voluntarism, although both are often thought of as proponents of that doctrine.

I am somewhat sceptical that anyone has actually followed such a procedure in order to acquire a belief. Nevertheless, it does seem possible. The important thing to note about this method of belief acquisition is that rather than giving support to the view that beliefs may be uncaused, it relies on the fact that beliefs are causally influenced by our environment or way of life. What is chosen is a course of action. It is then up to external causes to form the desired belief.

An example of a more common way of indirectly choosing to believe is offered by R.A. Sharpe (1975, pp. 213-14) in support of belief-voluntarism and as a counterexample to causal theories of knowledge. I could decide to find out what Mill said about the syllogism and then proceed to look up the appropriate books. If successful, I would come to know what Mill said about the syllogism. Since knowledge entails belief, I would come to believe that Mill said 'such-and-such' about the syllogism. Thus, I would have chosen to believe a specific proposition. Similarly, Roman Catholics may choose to believe everything that the Pope says *ex cathedra*.

But once again it is a case of choosing to place oneself in a certain epistemic situation; it is then up to that situation to give rise to the beliefs. Given the situation, the will has no further control over what those beliefs will be. If the authoritative book on Mill contains a misprint, my reaction may well be that I do not believe that that is what Mill said. No matter how devout a catholic may be, there are possible statements from the Pope that she will not believe. The Pope may go crazy and claim, for example, that white is black. Ignatius de Loyola asserted that 'we should always be ready to accept this principle: I will believe that the white that I see is black, if the hierarchical Church so defines it' (de Nicolas 1986, p. 172). But even for those who accept this principle, there is no guarantee that the required belief will follow. This underlines the fact that it is the epistemic situation including beliefs that have not been chosen, directly or indirectly, which controls what belief will be formed.

Suppose we do discover a method that is guaranteed to produce a given belief. Perhaps it involves using hypnosis or a 'credulity' drug. Choose the proposition *p* you would like to believe, swallow the drug, and (bingo!) you believe that *p*. Nobody has made a credulity pill. Yet! If this sort of method is the only possibility for our choosing beliefs then no one has ever chosen a belief and the belief-voluntarist position is severely restricted. And if a credulity pill were invented, it would pose no threat to a causal condition on knowledge. Such beliefs would have been caused in a certain way. The decision to take the drug or undergo the hypnosis would set in train a causal process culminating in the acquisition of the desired belief. It would simply remain for the particular causal theory to adjudicate as to whether that would be an 'appropriate' way for the belief to be caused if it is to qualify as knowledge.

Here is another example of indirectly choosing to believe. I would like to believe that there is a raven on the writing desk, so I place a raven on the writing desk. This action causes me to believe that there is a raven on the writing desk. Not only have I chosen my belief, but I now know that there is a raven on the writing desk. We can make such examples even more direct. I would like to believe that my arm is raised, so I raise my arm. In so doing, I come to believe that my arm is raised. But although this is as clear a case of choosing to believe as we have met, it is a case of indirectly choosing. My choice causes the action, but it is the action or a result of that action which causes my belief. I mention these examples because it might be suggested that these are cases where my control is as direct as it is in cases of bodily movement. Donald Davidson (1963) has a theory of events according to which (to give his example) the event of my flicking the switch is the same event as that of my turning on the light, which is the same event as my alerting the burglar. On this view it would seem that my alerting the burglar is as directly under my control as my flicking the switch, since they are simply the same event under different descriptions. Similarly, if I start a race by raising my arm, then my starting the race would be as directly under my control as my raising my arm. So why should not my believing that my arm is raised also be as directly under my control as my raising my arm? Simply because, even on the Davidsonian view, my coming to believe my arm is raised is not the same event as my raising my arm. According to Davidson, events are identical if and only if they have the same causes and effects. But my raising my arm is the cause of my believing that my arm is raised, and my raising my arm has other effects, such as the start of the race, which are not effects of my believing.

These examples of freely chosen actions causing beliefs about those actions are in fact excellent examples of beliefs that meet a causal requirement. In particular, the strong causal condition (SC) is met since the fact that my arm is raised causes my belief that my arm is raised.

It might be argued that the causal processes that give rise to belief in any of the above examples of indirect belief choice contain at least one event that is uncaused, namely, the decision to bring about the appropriate process, and that this would be enough to threaten the causal requirement. But if the universe was caused by an uncaused event (e.g. the Big Bang), then, according to this view, nothing would qualify as being caused in an appropriate way. Every event would have an uncaused event as part of its causal ancestry. I say more about events that are partially uncaused in Section 3.2.

William Alston (1989) identifies yet another possible way in which beliefs could be voluntarily chosen—one that comes between the indirect methods I have just been discussing and the direct choosing to believe I discussed earlier. Alston calls it 'nonbasic immediate control' (pp. 127-33).[3] If I raise my arm at will that is an example of what he calls 'basic control'. But if I open a door at will (by pushing it with my hand) then that is 'nonbasic immediate control'. It is 'immediate' control because, although more is required than just the movement of my hand, the action is performed in one uninterrupted intentional act. On the other hand, if I persuade someone else to open the door, then this would be an example of indirectly opening the door at will (what Alston calls 'long-range voluntary' control). Do we have nonbasic immediate control of our beliefs? If we did, what would it be like? We would have the ability to perform some (presumably mental) act that would have the immediate consequence that we form a belief. But we simply do not have such an ability. Imagining is a mental state or action over which we do have control. We can decide to imagine that the sun is shining, or that God exists, or that the left-hand track in the bush will lead to safety. But there is no special way of imagining that will automatically give rise to our believing that the sun is shining, or that God exists, or that the track leads to safety. Sometimes, such an imaginative act may be sufficient (when added to all other influences) to produce a belief, but on other occasions it may have the opposite effect. Clearly imagining what the world would be like if God exists may help to produce a genuine belief in God, but on the other hand, it may be enough for someone to realise the foolishness of such an hypothesis. In either case, it will be what seems to us at the time to be true which will be crucial, and we have no control over that. Alternatively, imagining something to be the case may help us to more

---

[3] Alston also identifies a fourth possibility, 'indirect influence' (p. 136), which I do not discuss because it is more relevant to his main concern which is the deontological conception of epistemic justification

effectively behave as if we believed it to be true. Imagining that I am on the path to safety may help me to continue on that path with greater resolution.

There is no other mental act or state subject to our will that would have the desired effect of producing a specified belief. We just do not (as Alston puts it) 'know which mental button to push'. Future psychologists may uncover sufficient secrets of the human mind to enable them to teach us how it can be done, perhaps with the aid of a little neurosurgical tinkering. But at the moment, it is a contingent fact that we lack the ability. Even if we did acquire such an ability, the amendment I suggest in Section 3.3 avoids any threat to a causal condition on knowledge.

To sum up, we cannot directly choose our beliefs. If we have freewill, then that is the freedom to choose our actions not our beliefs. Any process by which we seem to be directly choosing to believe turns out either to be a process by which we discover what we believe by discovering what seems to us to be the case, or else to be a process by which we choose to act as though we had the belief. Any process by which we indirectly choose to believe turns out to be a process by which we choose to act so as to expose ourselves to causal influences that may give rise to a particular belief. Such processes are compatible with a causal requirement on knowledge.

BIBLIOGRAPHY

Alston, W.P. (1989) *Epistemic Justification*. Ithaca: Cornell University Press.

Anderson, C.A. (1993) 'Zalta's Intensional Logic.' *Philosophical Studies* 69: 221-29.

Armstrong, D.M. (1969-70) 'Does Knowledge Entail Belief?' *Proceedings of the Aristotelian Society* 70: 21-36.

Armstrong, D.M. (1973) *Belief, Truth and Knowledge.* Cambridge: Cambridge University Press.

Armstrong, D.M. (1978a) *Nominalism and Realism,* vol. 1 of *Universals and Scientific Realism.* Cambridge: Cambridge University Press.

Armstrong, D.M. (1978b) *A Theory of Universals,* vol. 2 of *Universals and Scientific Realism.* Cambridge: Cambridge University Press.

Armstrong, D.M. (1983) *What is a Law of Nature?* Cambridge: Cambridge University Press.

Azzouni, J. (1994) *Metaphysical Myths, Mathematical Practice: The Ontology and Epistemology of the Exact Sciences.* Cambridge: Cambridge University Press.

Bacon, F. (1620/1960), 'Novum Organum', in *The New Organum and Related Writing*, edited by F.H. Anderson. New York: Liberal Arts Press.

Balaguer, M. (1995) 'A Platonist Epistemology.' *Synthese* 103: 303-25.

Balaguer, M. (1998a) 'Non-Uniqueness as a Non-Problem', *Philosophia Mathematica* (III) 6: 63-84.

Balaguer, M. (1998b) *Platonism and Anti-Platonism in Mathematics.* New York: Oxford University Press.

Barwise, J. & Perry, J. (1983) *Situations and Attitudes.* Cambridge, MA: The MIT Press.

Bastick, T. (1982) *Intuition: How we think and act.* Chichester: J. Wiley & Sons.

Beall, JC. (2001) 'Existential Claims and Platonism.' *Philosophia Mathematica* (III) 9: 80-86

Benacerraf, P. (1973) 'Mathematical Truth.' in Benacerraf & Putnam (eds) (1983): 403-20.

Benacerraf, P. & Putnam, H. (eds) (1983) *Philosophy of Mathematics,* 2nd edn. Cambridge: Cambridge University Press.

Bennett, J. (1990) 'Why is Belief Involuntary?' *Analysis* 50: 87-107.

Berkeley, G. (1710/1965) *The Principles of Human Knowledge*, in *Berkeley's Philosophical Writings*, edited by D.M. Armstrong. New York: Collier Books.

Bigelow, J. (1988) *The Reality of Numbers.* Oxford: Clarendon Press.

Bigelow, J. (1990) 'Sets Are Universals.' in Irvine (ed.) (1990): 291-305.

Bloor, D. (1976) *Knowledge and Social Imagery.* London: Routledge & Kegan Paul.

Brown, J.R. (1989) *The Rational and the Social.* London: Routledge.

Brown, J.R. (1990) 'π in the Sky.' in Irvine (ed.) (1990): 95-120.

Burgess, J.P. (1990) 'Epistemology and Nominalism.' in Irvine (ed.) (1990): 1-15.

Burgess, J.P. & Rosen, G. (1997) *A Subject with No Object.* Oxford: Clarendon Press.

Campbell, K. (1990) *Abstract Particulars.* Oxford: Blackwell.

Carrier, L.S. (1976) 'The Causal Theory of Knowledge.' *Philosophia* 6: 237-57.

Cartwright, N. (1983) *How the Laws of Physics Lie.* Oxford: Clarendon Press.

Casullo, A. (1992) 'Causality, Reliabilism, and Mathematical Knowledge.' *Philosophy and Phenomenological Research* 52: 557-84.

Chang, H. & Cartwright, N. (1993) 'Causality and Realism in the EPR Experiment.' *Erkenntnis* 38: 169-90.

Cheyne, C. (1989) *The Viability of Mathematical Empiricism.* Unpublished DipArts dissertation, University of Otago.

Cheyne, C. (1994) *Knowledge, Cause and Platonic Objects.* Unpublished PhD thesis, University of Otago.

Cheyne, C. (1997a) 'Epistemic Value and Fortuitous Truth.' *Principia: revista internacional de epistemologia* 1: 109-34.

Cheyne, C. (1997b) 'Getting in Touch with Numbers: Intuition and Mathematical Platonism.' *Philosophy and Phenomenological Research* 57: 111-25.

Cheyne, C. (1998) 'Existence Claims and Causality.' *Australasian Journal of Philosophy* 76: 34-47.

Cheyne, C. (1999) 'Problems with Profligate Platonism.' *Philosophia Mathematica* (III) 7: 164-77.

Cheyne, C. & Pigden, C. (1996) 'Pythagorean Powers or A Challenge to Platonism.' *Australasian Journal of Philosophy* 74: 639-45.

Chihara, C. (1982) 'A Gödelian Thesis Regarding Mathematical Objects: Do they Exist? And can we Perceive them?' *Philosophical Review* 91: 211-27.

Chihara, C. (1990) *Constructibility and Mathematical Existence.* Oxford: Clarendon Press.

Churchland, P.M. (1981) 'Eliminative Materialism and Propositional Attitudes.' *Journal of Philosophy* 78: 67-90.

Close, F. & Maddox, J. (1994) 'Getting to the Bottom of the Top.' *Nature* 368: 805.

Cohen, P.J. (1966) *Set Theory and the Continuum Hypothesis.* New York: W.A. Benjamin.

Colman, A.M. (1987) *Facts, Fallacies, and Frauds in Psychology.* London: Unwin Hyman.

Colyvan, M. (1998a) 'Can the Eleatic Principle Be Justified?' *Canadian Journal of Philosophy* 28: 313-35.

Colyvan, M. (1998b) 'Is Platonism a Bad Bet?' *Australasian Journal of Philosophy* 76: 115-19.

d'Espagnat, B. (1979) 'The Quantum Theory and Reality.' *Scientific American* 24: 158-81.

Davidson, D. (1963) 'Actions, Reasons and Causes.' *Journal of Philosophy* 60: 685-700.

Davis, P.J. & Hersh, R. (1981) *The Mathematical Experience.* Boston: Birkhäuser.

de Nicolas, A.T. (1986) *Powers of Imagining: Ignatius de Loyola.* Albany, NY: SUNY Press.

Descartes, R. (1641/1969) *Meditations on First Philosophy* in Wilson (1969).

Descartes, R. (1649/1969) *The Passions of the Soul* in Wilson (1969).

Deutsch, H. (1993) 'Zalta on Sense and Substitutivity.' *Philosophical Studies* 69: 209-19.

Devitt, M. (1980) '"Ostrich Nominalism" or "Mirage Realism"?' *Pacific Philosophical Quarterly* 61: 433-39.

Devitt, M. (1981) *Designation.* New York: Columbia University Press.

Devitt, M. & Sterelny, K. (1987) *Language and Reality.* Oxford: Blackwell.

Dretske, F. (1977a) 'Perception and Reference without Causality.' *Journal of Philosophy* 74: 621-25.

Dretske, F. (1977b) 'Laws of Nature.' *Philosophy of Science* 44: 248-68.

Dretske, F. & Enç, B. (1984) 'Causal Theories of Knowledge.' in French et al. (eds) (1984): 517-28.

Feldman, R. (1974) 'An Alleged Defect in Gettier-Counterexamples.' *Australasian Journal of Philosophy* 52: 68-69.

Field, H. (1980) *Science Without Numbers.* Oxford: Blackwell.

Field, H. (1989) *Realism, Mathematics and Modality.* Oxford: Blackwell.

Fine, K. (1975) 'Vagueness, Truth and Logic.' *Synthese* 30: 265-300.

Fodor, J.A. (1976) *The Language of Thought.* Sussex: The Harvester Press.

Folina, J. (2000) 'Ontology, Logic, and Mathematics.' *The British Journal for the Philosophy of Science* 51: 319-32.

Frankfurt, H. (1971) 'Freedom of the Will and the Concept of a Person.' *Journal of Philosophy* 68: 5-20.

Frege, G. (1960) *Translations from the Philosophical Writings of Gottlob Frege,* 2nd edn. edited by P. Geach and M. Black. Oxford: Blackwell.

Frege, G. (1964) *The Basic Laws of Arithmetic.* transl. by M. Fürth. Berkeley & Los Angeles: University of California Press.

Frege, G. (1968) *The Foundations of Arithmetic,* transl. by J.L. Austin, 2nd edn. Oxford: Blackwell.

French, P.A. et al. (eds) (1984) *Causation and Causal Theories: Midwest Studies in Philosophy IX.* Minneapolis: University of Minnesota Press.

Frické, M. (ed.) (1986) *Essays in Honour of Bob Durrant.* Dunedin: Otago University Philosophy Department.

Fumerton, R.A. (1987) 'Nozick's Epistemology.' in Luper-Foy (ed.) (1987): 163-81.

Gettier, E.L. (1963) 'Is Justified True Belief Knowledge?' *Analysis,* 23: 121-23.

Gibbon, E. (1776-88/1994) *The History of the Decline and Fall of the Roman Empire*, edited by D. Womersley. London: Allen Lane.

Ginet, C. (1975) *Knowledge, Perception, and Memory.* Dordrecht: D. Reidel.

Gödel, K. (1947/1983) 'What is Cantor's Continuum Problem?' in Benacerraf & Putnam (1983): 470-85.

Goldman, A.H. (1988) *Empirical Knowledge.* Berkeley: University of California Press.

Goldman, A.I. (1967) 'A Causal Theory of Knowing.' *Journal of Philosophy* 64: 357-72.

Goldman, A.I. (1976) 'Discrimination and Perceptual Knowledge.' *Journal of Philosophy* 73: 771-91.

Goldman, A.I. (1986) *Epistemology and Cognition.* Cambridge, MA: Harvard University Press.

Goldman, A.I. (1989) 'Psychology and Philosophical Analysis.' *Proceedings of the Aristotelian Society* 89: 195-209.

Goldstick, D. (1972) 'A Contribution towards the Development of a Causal Theory of Knowledge.' *Australasian Journal of Philosophy* 50: 238-48.

Govier, T. (1976) 'Beliefs, Values, and the Will.' *Dialogue* 15: 642-63.

Grice, H.P. (1957) 'Meaning.' *Philosophical Review* 66: 377-88.

Grice, H.P. (1961) 'The Causal Theory of Perception.' in Swartz (1961): 438-72.

Grosser, M. (1962) *The Discovery of Neptune.* Cambridge, MA: Harvard University Press.

Hacking, I. (1983) *Representing and Intervening: Introductory Topics in the Philosophy of Natural Science.* Cambridge: Cambridge University Press.

Hale, B. (1987) *Abstract Objects.* Oxford: Blackwell.

Hale, B. (1990) 'Nominalism.' in Irvine (1990): 121-44.

Hale, B. (1994) 'Is Platonism Epistemologically Bankrupt?' *Philosophical Review* 103: 299-325.

Hale, B. & Wright, C. (1992) 'Nominalism and the Contingency of Abstract Objects.' *Journal of Philosophy* 89: 111-35.

Harman, G. (1973) *Thought.* Princeton: Princeton University Press.

Hart, W.D. (1977) 'Review of Steiner's *Mathematical Knowledge.*' *Journal of Philosophy* 74: 118-29.

Hellman, G. (1989) *Mathematics without Numbers.* Oxford: Clarendon Press.

Hofstadter, D. (1979) *Gödel, Escher, Bach: an Eternal Golden Braid.* Harmondsworth: Penguin Books.

Holton, G. & Roller, D. (1958) *Foundations of Modern Physical Science.* Reading, MA: Addison-Wesley.

Horwich, P. (1991) 'On the Nature and Norms of Theoretical Commitment.' *Philosophy of Science* 58: 1-14.

Hughes, G.E. & Cresswell, M.J. (1968) *An Introduction to Modal Logic.* London: Methuen & Co.

Hume, D. (1739-40/1978) *A Treatise of Human Nature.* edited. by L. A. Selby-Bigge. 2nd edn. Oxford: Oxford University Press.

Irvine, A.D. (ed.) (1990) *Physicalism in Mathematics.* Dordrecht: Kluwer Academic Publishers.

James, W. (1897/1979) *The Will to Believe and Other Essays in Popular Philosophy*, edited by F. Burkhardt, et al. Cambridge, MA: Harvard University Press.

Kant, I. (1781/1929) *Critique of Pure Reason*, transl. by N. Kemp Smith). London: Macmillan & Co.

Katz, J.J. (1998) *Realistic Rationalism.* Cambridge, MA: The MIT Press.

Kim, J. (1975) 'Events as Property Exemplifications.' in *Action Theory*, edited by M. Brand & D. Walton. Dordrecht: Reidel.

Kim, J. (1977) 'Perception and Reference without Causality.' *Journal of Philosophy* 74: 606-20.

Kitcher, P. (1984) *The Nature of Mathematical Knowledge.* Oxford: Oxford University Press.

Klein, P.D. (1976) 'Knowledge, Causality, and Defeasibility.' *Journal of Philosophy* 68: 792-812.

Kripke, S. (1980) *Naming and Necessity.* Oxford: Blackwell.

Kurtz, P. (ed.) (1985) *The Skeptic's Handbook of Parapsychology.* Buffalo: Prometheus Books.

Lehrer, K. (1965) 'Knowledge, Truth and Evidence.' *Analysis* 25: 168-75.

Lehrer, K. (1974) *Knowledge.* Oxford: Clarendon Press.

Lehrer, K. (1990) *Theory of Knowledge.* London: Routledge.

Lehrer, K. & Paxson, T.D. (1969) 'Knowledge: Undefeated Justified True Belief.' in Pappas & Swain (eds) (1978): 146-54.

Lewis, D.K. (1986) *On the Plurality of Worlds.* Oxford: Blackwell.

Lewis, D.K. (1992) 'Critical Notice of Armstrong's *A Combinatorial Theory of Possibility*.' *Australasian Journal of Philosophy* 70: 211-24.

Liston, M. (1993) 'Taking Mathematical Fictions Seriously.' *Synthese* 95: 433-58.

Linsky, B. & Zalta, E. (1995) 'Naturalized Platonism versus Platonized Naturalism.' *Journal of Philosophy* 92: 525-55.

Locke, J. (1690/1976) *An Essay Concerning Human Understanding*, abridged and edited by J.W. Yolton. London: Dent.

Loeb, L.E. (1976) 'On a Heady Attempt to Befiend Causal Theories of Knowledge.' *Philosophical Studies* 29: 331-36.

Luper-Foy, S. (ed.) (1987) *The Possibility of Knowledge: Nozick and his Critics.* Totowa: Rowman & Littlefield.

Lycan, W.G. (1977) 'Evidence One Does Not Possess.' *Australasian Journal of Philosophy* 55: 114-26.

Lycan, W.G. (1988) *Judgement and Justification.* Cambridge: Cambridge University Press.

Mackie, J. (1974) *The Cement of the Universe.* Oxford: Oxford University Press.

Mac Lane, S. (1986) *Mathematics, Form and Function.* New York: Springer-Verlag.

Maddy, P. (1980) 'Perception and Mathematical Intuition.' *Philosophical Review* 89: 163-96.

Maddy, P. (1984) 'Mathematical Epistemology: What is the Question?' *Monist* 67: 46-55.

Maddy, P. (1990) *Realism in Mathematics.* Oxford: Clarendon Press.

Maddy, P. (1992) 'Indispensability and Practice.' *Journal of Philosophy* 89: 275-89.

Malcolm, N. (1960) 'Anselm's Ontological Arguments.' *Philosophical Review* 69: 41-62.

Mannison, D.S. (1976) '"Inexplicable Knowledge" Does Not Require Belief.' *Philosophical Quarterly* 26: 139-48.

McFetridge, I.G. (1985) 'Review of Philip Kitcher's *The Nature of Mathematical Knowledge.*' *Mind* 94: 321-23.

Mellor, D.H. (1995) *The Facts of Causation.* London: Routledge.

Menzel, C. (1993) 'Possibilism and Object Theory.' *Philosophical Studies* 69: 195-208.

Menzies, P. (1989) 'A Unified Account of Causal Relata.' *Australasian Journal of Philosophy*, 67: 59-83.

Meyers, R. & Stern, K. (1973) 'Knowledge without Paradox.' *Journal of Philosophy* 70: 147-60.

Mill, J.S. (1843/1973) *A System of Logic, Raciocinative and Inductive.* London: Routledge & Kegan Paul.

Mill, J.S. (1873/1992) *The Autobiography of John Stuart Mill,* edited by A.O.J. Cockshut. Halifax: Ryburn Publishing.

Miller, D. (1987) 'A Critique of Good Reasons.' in *Rationality: The Critical View,* edited by J. Agassi & I. Jarvie. Dordrecht: Martinus Nijhoff.

Musgrave, A. (1986) 'Arithmetical Platonism: Is Wright Wrong or Must Field Yield?' in Frické (ed.) (1986): 90-110.

Musgrave, A. (1989) 'Deductivism versus Psychologism.' in *Perspectives on Psychologism,* edited. by M.A. Notturno. Leiden: E.J. Brill.

Musgrave, A. (1991) 'The Myth of Astronomical Instrumentalism.' in *Beyond Reason: Essays on the Philosophy of Paul Feyerabend,* edited by G. Munevar. Dordrecht: Kluwer Academic Publishers.

Musgrave, A. (1993) *Common Sense, Science and Scepticism.* Cambridge: Cambridge University Press.

Mundy, B. (1992) 'Space-Time and Isomorphism.' in *PSA 1992: Proceedings of the 1992 Biennial Meeting of the Philosophy of Science Association* vol. 1, edited by D. Hull et al. East Lansing: Philosophy of Science Association.

Noddings, N. & Shore, P. (1984) *Awakening the Inner Eye: Intuition in Education.* New York: Teachers College Press.

Nozick, R. (1981) *Philosophical Explanations.* Cambridge, MA: Harvard University Press.

Nye, M.J. (1972) *Molecular Reality: A Perspective on the Scientific Work of Jean Perrin.* London: Macdonald & Co.

Papineau, D. (1988) 'Mathematical Fictionalism.' *International Studies in the Philosophy of Science* 2: 151-74.

Pappas, G. & Swain, M. (eds) (1978) *Essays on Knowledge and Justification.* Ithaca, NY: Cornell University Press.

Parsons, C. (1979-80) 'Mathematical Intuition.' *Proceedings of the Aristotelian Society* 80: 145-68.

Parsons, C. (1983) *Mathematics in Philosophy: Selected Essays.* Ithaca: Cornell University Press.

Pascal, B. (1670/1966) *Pensées,* transl. by A.J. Krailsheimer. Harmondsworth: Penguin.

Paton, H.J. (1948) *The Moral Law: Kant's 'Groundwork of the Metaphysic of Morals'*. London: Hutchinson.

Plantinga, A. (1974) *The Nature of Necessity*. Oxford: Oxford University Press.

Plantinga, A. (1988) 'Positive Epistemic Status and Proper Function.' in *Philosophical Perspectives 2, Epistemology*, edited by J. Tomberlin. Atascadero: Ridgeway Publishing Company.

Plantinga, A. (1990) 'Justification in the 20th Century.' *Philosophy and Phenomenological Research* 50: 45-71.

Plantinga, A. (1993) *Warrant and Proper Function*. Oxford: Oxford University Press.

Plato (1892) *The Dialogues of Plato* 3rd edn. transl. by B. Jowett in 5 volumes. London: Oxford University Press.

Price, H.H. (1961) 'The Causal Theory.' in Swartz (1961): 394-437.

Putnam, H. (1979) 'Philosophy of Logic' in *Mathematics, Matter and Method*, 2nd edn. Cambridge: Cambridge University Press.

Quine, W.V.O. (1961) *From a Logical Point of View*, 2nd edn. New York: Harper & Row.

Quine, W.V.O. (1966) *The Ways of Paradox*. New York: Random House.

Quine, W.V.O. (1969) 'Epistemology Naturalized.' in his *Ontological Relativity and Other Essays*. New York: Columbia University Press.

Radford, C. (1966) 'Knowledge—By Examples.' *Analysis* 27: 1-11.

Resnik, M.D. (1990) 'Beliefs About Mathematical Objects.' in Irvine (ed.) (1990): 41-71.

Resnik, M.D. (1991) 'Between Mathematics and Physics' in *PSA 1990: Proceedings of the 1990 Biennial Meeting of the Philosophy of Science Association* vol. 2, edited by A. Fine et al. East Lansing: Philosophy of Science Association.

Ruben, D.H. (1990) *Explaining Explanation*. London: Routledge.

Russell, B. (1919) *Introduction to Mathematical Philosophy*. London: George Allen and Unwin.

Saunders, J.T. & Champawat, N. (1964) 'Mr. Clark's Definition of "Knowledge".' *Analysis* 25: 8-9.

Sextus Empiricus (1936) 'Against the Logicians II.' in *Sextus Empiricus* vol. 2, transl. by R.G. Bury. London: Heinemann.

Shapiro, S. (1983) 'Conservativeness and Incompleteness.' *Journal of Philosophy* 80: 521-31.

Shapiro, S. (1993) 'Modality and Ontology.' *Mind* 102: 455-81.

Shapiro, S. (1997) *Philosophy of Mathematics: Structure and Ontology.* Oxford: Oxford University Press.

Sharpe, R.A. (1975) 'On the Causal Theory of Knowledge.' *Ratio* 17: 206-16.

Shope, R.K. (1983) *The Analysis of Knowing.* Princeton: Princeton University Press.

Skyrms, F.B. (1967) 'The Explication of "X knows that *p*".' *Journal of Philosophy* 64: 373-89.

Slater, B.H. (1997) 'De-Mystifying Situations.' *Philosophical Papers* 26: 165-78.

Sober, E. (1993) 'Mathematics and Indispensability.' *The Philosophical Review* 102: 35-57.

Staniland, H. (1972) *Universals.* London: Macmillan.

Steiner, M. (1975) *Mathematical Knowledge.* Ithaca: Cornell University Press.

Stove, D. (1972) 'Misconditionalisation.' *Australasian Journal of Philosophy* 50: 173-83.

Stove, D. (1991) *The Plato Cult and Other Philosophical Follies.* Oxford: Blackwell.

Swartz, R. (ed.) (1961) *Perceiving, Sensing and Knowing.* Berkeley: University of California Press.

Tarski, A. (1983) 'On the Concept of Logical Consequence.' in his *Logic, Semantics, Metamathematics*, 2nd edn., transl. by J.H. Woodger. Indianapolis: Hackett Publishing Company.

Taylor, B. (1985) *Modes of Occurrence.* Oxford: Blackwell.

Taylor, J.E. (1991) 'Plantinga's Proper Functioning Analysis of Epistemic Warrant.' *Philosophical Studies* 64: 185-202.

Thalberg, I. (1969) 'In Defense of Justified True Belief.' *Journal of Philosophy* 66: 795-803.

Tieszen, R.L. (1989) *Mathematical Intuition: Phenomenology and Mathematical Knowledge.* Dordrecht: Kluwer.

Tooley, M. (1977) 'The Nature of Laws.' *Canadian Journal of Philosophy* 7: 667-98.

van Fraassen, B. (1969) 'Presuppositions, Supervaluations, and Free Logic.' in *The Logical Way of Doing Things*, edited by K. Lambert. New Haven: Yale University Press.

van Fraassen, B. (1980) *The Scientific Image.* Oxford: Clarendon Press.

Wedberg, A. (1955) *Plato's Philosophy of Mathematics.* Stockholm: Almquist & Wiksell.

Williams, B. (1972) 'Deciding to Believe.' in his *Problems of the Self.* Cambridge: Cambridge University Press.

Williams, D.C. (1966) 'The Elements of Being.' in his *The Principles of Empirical Realism.* Springfield: Charles Thomas.

Wilson, M.D. (ed.) (1969) *The Essential Descartes.* New York: New American Library.

Winters, B. (1979) 'Believing at Will.' *Journal of Philosophy* 76: 243-56.

Wright, C. (1983) *Frege's Conception of Numbers as Objects.* Aberdeen: Aberdeen University Press.

Wright, C. (1988) 'Why Numbers Can Believably Be.' *Revue Internationale de Philosophie* 42: 425-73.

Zalta, E. (1983) *Abstract Objects: An Introduction to Axiomatic Metaphysics.* Dordrecht: D. Reidel.

Zalta, E. (1988) *Intensional Logic and the Metaphysics of Intentionality* Cambridge, MA: The MIT Press

Zalta, E. (1993) 'Replies to the Critics.' *Philosophical Studies* 69: 231-42.

INDEX

The Western Ontario Series
in Philosophy of Science

1. J. Leach, R. Butts and G. Pearce (eds.): *Science, Decision and Value.* 1973
    ISBN 90-277-0239-X; Pb 90-277-0327-2

2. C.A. Hooker (ed.): *Contemporary Research in the Foundations and Philosophy of Quantum Theory.* 1973   ISBN 90-277-0271-3; Pb 90-277-0338-8

3. J. Bub: *The Interpretation of Quantum Mechanics.* 1974
    ISBN 90-277-0465-1; Pb 90-277-0466-X

4. D. Hockney, W. Harper and B. Freed (eds.): *Contemporary Research in Philosophical Logic and Linguistic Semantics.* 1975   ISBN 90-277-0511-9; Pb 90-277-0512-7

5. C.A. Hooker (ed.): *The Logico-algebraic Approach to Quantum Mechanics.*
    Vol. I: Historical Evolution. 1975   ISBN 90-277-0567-4; Pb 90-277-0613-1
    Vol. II: Contemporary Consolidation. 1979   ISBN 90-277-0707-3; Pb 90-277-0709-X

6. W.L. Harper and C.A. Hooker (eds.): *Foundations of Probability Theory, Statistical Inference, and Statistical Theories of Science.*
    Vol. I: Foundations and Philosophy of Epistemic Applications of Probability Theory. 1976
    ISBN 90-277-0616-6; Pb 90-277-0617-4
    Vol. II: Foundations and Philosophy of Statistical Inference. 1976
    ISBN 90-277-0618-2; Pb 90-277-0619-0
    Vol. III: Foundations and Philosophy of Statistical Theories in the Physical Sciences. 1976
    ISBN 90-277-0620-4; Pb 90-277-0621-2

7. C.A. Hooker (ed.): *Physical Theory as Logico-operational Structure.* 1979
    ISBN 90-277-0711-1

8. J.M. Nicholas (ed.): *Images, Perception, and Knowledge.* 1977   ISBN 90-277-0782-0

9. R.E. Butts and J. Hintikka (eds.): *Logic, Foundations of Mathematics, and Computability Theory.* Part One: Logic, Foundations of Mathematics, and Computability Theory. 1977
    ISBN 90-277-0708-1

10. R.E. Butts and J. Hintikka (eds.): *Logic, Foundations of Mathematics, and Computating Theory.* Part Two: Foundational Problems in the Special Sciences. 1977   ISBN 90-277-0710-3

11. R.E. Butts and J. Hintikka (eds.): *Logic, Foundations of Mathematics, and Computability Theory.* Part Three: Basic Problems in Methodology and Linguistics. 1977
    ISBN 90-277-0829-0

12. R.E. Butts and J. Hintikka (eds.): *Logic, Foundations of Mathematics, and Computability Theory.* Part Four: Historical and Philosophical Dimensions of Logic, Methodology and Philosophy of Science. 1977   ISBN 90-277-0831-2
    Set (9-12) ISBN 90-277-0706-5

13. C.A. Hooker J.J. Leach and E.F. McClennen (eds.): *Foundations and Applications of Decision Theory.*
    Vol. I: Theoretical Foundations. 1978   ISBN 90-277-0842-8
    Vol. II: Epistemic and Social Applications. 1978   ISBN 90-277-0844-4

14. R.E. Butts and J.C. Pitt (eds.): *New Perspectives on Galileo.* 1978
    ISBN 90-277-0859-2; Pb 90-277-0891-6

15. W.L. Harper, R. Stalnaker and G. Pearce (eds.): *Ifs. Conditionals, Belief, Decision, Chance, and Time.* 1981   ISBN 90-277-1184-4; Pb 90-277-1220-4

16. J.C. Pitt (ed.): *Philosophy in Economics.* 1981   ISBN 90-277-1210-7; Pb 90-277-1242-5

The Western Ontario Series
in Philosophy of Science

The Western Ontario Series
in Philosophy of Science

39. I.B. MacNeill and G.J. Umphrey (eds.): *Advances in the Statistical Sciences.*
     Vol. VI: Actuarial Science. 1987   ISBN 90-277-2398-2
                                        Set ISBN (Vols 34-39) 90-277-2399-0
40. N. Rescher: *Scientific Realism.* A Critical Reappraisal. 1987
                                        ISBN 90-277-2442-3; Pb 90-277-2528-4
41. B. Skyrms and W.L. Harper (eds.): *Causation, Chance, and Credence.* 1988
                                                          ISBN 90-277-2633-7
42. W.L. Harper and B. Skyrms (eds.): *Causation in Decision, Belief Change and Statistics.* 1988
                                                          ISBN 90-277-2634-5
43. R.S. Woolhouse (ed.): *Metaphysics and Philosophy of Science in the 17th and 18th Centuries.*
     Essays in Honor of Gerd Buchdahl. 1988   ISBN 90-277-2743-0
44. R.E. Butts and J.R. Brown (eds.): *Constructivism and Science.* Essays in Recent German
     Philosophy. 1989   ISBN 0-7923-0251-6
45. A.D. Irvine (ed.): *Physicalism in Mathematics.* 1989   ISBN 0-7923-0513-2
46. J. van Cleve and R.E. Frederick (eds.): *The Philosophy of Right and Left.* Incongruent Coun-
     terparts and the Nature of Space. 1991   ISBN 0-7923-0844-1
47. F. Wilson: *Empiricism and Darwin's Science.* 1991   ISBN 0-7923-1019-5
48. G.G. Brittan, Jr. (ed.): *Causality, Method and Modality.* Essays in Honor of Jules Vuillemin.
     1991   ISBN 0-7923-1045-4
49. W. Spohn, B.C. van Fraassen and B. Skyrms (eds.): *Existence and Explanation.* Essays in
     Honor of Karel Lambert. 1991   ISBN 0-7923-1252-X
50. J.C. Pitt: *Galileo, Human Knowledge, and the Book of Nature.* Method Replaces Metaphysics.
     1992   ISBN 0-7923-1510-3
51. V. Coelho (ed.): *Music and Science in the Age of Galileo.* 1992   ISBN 0-7923-2028-X
52. P. Janich: *Euclid's Heritage: Is Space Three-Dimensional?* 1992   ISBN 0-7923-2025-5
53. M. Carrier: *The Completeness of Scientific Theories.* On the Derivation of Empirical Indicators
     within a Theoretical Framework: The Case of Physical Geometry. 1994
                                                          ISBN 0-7923-2475-7
54. P. Parrini (ed.): *Kant and Contemporary Epistemology.* 1994   ISBN 0-7923-2681-4
55. J. Leplin (ed.): *The Creation of Ideas in Physics.* Studies for a Methodology of Theory
     Construction. 1995   ISBN 0-7923-3461-2
56. J.E. McGuire: *Tradition and Innovation.* Newton's Metaphysics of Nature. 1995
                                                          ISBN 0-7923-3617-8
57. R. Clifton (ed.): *Perspectives on Quantum Reality.* Non-Relativistic, Relativistic, and Field-
     Theoretic. 1996   ISBN 0-7923-3812-X
58. P.H. Theerman and K. Hunger Parshall (eds.): *Experiencing Nature.* Proceedings of a Confer-
     ence in Honor of Allen G. Debus. 1997   ISBN 0-7923-4477-4
59. P. Parrini: *Knowledge and Reality.* An Essay in Positive Philosophy. 1998
                                                          ISBN 0-7923-4939-3
60. D. Dieks and P.E. Vermaas (eds.): *The Modal Interpretation of Quantum Mechanics.* 1998
                                                          ISBN 0-7923-5207-6

The Western Ontario Series
in Philosophy of Science

61. M.C. Galavotti and A. Pagnini (eds.): *Experience, Reality, and Scientific Explanation*. 1999
ISBN 0-7923-5497-4

62. D. Fisette (ed.): *Consciousness and Intentionality: Models and Modalities of Attribution*. 1999
ISBN 0-7923-5907-0

63. J.L. Bell: *The Art of the Intelligible*. An Elementary Survey of Mathematics in its Conceptual Development. 1999
ISBN 0-7923-5972-0

64. P. Gozza: *Number to Sound*. The Musical Way to the Scientific Revolution. 1999
ISBN 0-7923-6069-9

65. R.E. Butts: *Witches, Scientists, Philosophers: Essays and Lectures*. (Ed. by G. Solomon). 2000
ISBN 0-7923-6608-5

66. L. Magnani: *Philosophy and Geometry*. Theoretical and Historical Issues. 2001
ISBN 0-7923-6933-5

67. C. Cheyne: *Knowledge, Cause, and Abstract Objects*. Causal Objections to Platonism. 2001
ISBN 1-4020-0051-0

KLUWER ACADEMIC PUBLISHERS – DORDRECHT / BOSTON / LONDON